THE EMPLOYMENT OF ENGLISH

THE EMPLOYMENT
OF ENGLISH

THEORY, JOBS, AND THE FUTURE
OF LITERARY STUDIES

MICHAEL BÉRUBÉ

NEW YORK UNIVERSITY PRESS
NEW YORK AND LONDON

NEW YORK UNIVERSITY PRESS
New York and London

Library of Congress Cataloging-in-Publication Data
Bérubé, Michael, 1961–
The employment of English : theory, jobs, and the future of
literary studies / by Michael Bérubé.
p.   cm.—(Cultural front)
Includes bibliographical references and index.
ISBN 0-8147-1300-9 (pbk. : acid-free paper).—ISBN
0-8147-1301-7 (pbk. : acid-free paper)
1. English philology—Study and teaching (Higher)—Political
aspects—United States.   2. English literature—History and
criticism—Theory, etc.   3. English language—Political aspects—
United States.   4. English teachers—Employment—United States.
5. Interdisciplinary approach in education.   6. English philology—
Vocational guidance.   7. Language and culture—United States.
I. Title.   II. Series: Cultural front (Series)
PE68.U5B48   1988
820'.71'173—dc21   97-21213
                         CIP

New York University Press books are printed on acid-free paper,
and their binding materials are chosen for strength and durability.

Manufactured in the United States of America

10 9 8 7 6 5 4 3 2 1

CONTENTS

PREFACE

I love literature. I really do. And however much it might shock my colleagues, I believe that some books are better than others, too. Yet this belief of mine complicates my love: I like some books and some writers a great deal, whereas there are other books and other writers I don't care for at all. So now that I think of it, I guess it makes about as much sense to say "I love literature" as to say "I'm fond of food." It looks as if I'm going to have to start this preface all over again, this time from a coherent premise.

Then again, perhaps the incoherence of the premise should be my premise: how has it come to pass that in the profession of academic literary study, professors of English can publicly profess their love of "literature" (all of it, presumably) as if they were saying something meaningful? And loving literature as they do, renowned critics renounce criticism: some complain that academic criticism is too esoteric; some complain that *contemporary* academic criticism is too politicized; some complain that *any* kind of criticism inevitably positions itself as "superior" to its object, and is suspect on that basis alone. English departments throw themselves (and the very nation) into turmoil over whether to introduce a required "theory" course for graduate students—or whether to jettison the Chaucer-Milton-Shakespeare requirement for undergraduates. Conferences, colloquia, careers are devoted to asking whether literary criticism can have any productive political role in the world; conferences, colloquia, careers are devoted to asking whether literary criticism loves literature. Traditional defenses of the humanities no longer compel assent from students, trustees, or legislators (each for their own reasons more concerned with return on investment than with *Remembrance of Things Past*), and, accordingly, traditional defenses of the humanities are no longer attempted by junior faculty or graduate students. And year after year, thousands of new Ph.D.s in the field find that amid the myriad debates over the practices and prospects of the profession, there are no teaching jobs for them, regardless of what they think of the function of criticism at the present time.

What does the future look like for departments of English literature? Does academic literary study even *have* a future—or should it?

This book is about the intellectual and economic status of the profession of literary study at a time when "employment" and "English" are two of the most volatile and contested terms in the business. Some of the chapters that follow address the current employment conditions in English departments, by analyzing the professional tensions created by a substantially shrunken job market; other chapters address the social functions of literary study and interpretive theory, in English departments and in the broader culture. All of them were written in response to, and are informed by, the profession's competing (and, in some respects, mutually defining) fiscal and intellectual imperatives—the intersection of which, I argue, both constitutes and mystifies the crisis of reproduction in the modern languages. *The Employment of English* is thus divided, as its structure suggests, between attention to "employment *in* English" and the task of "employing English" (or the knowledges produced in English) outside the academy. I do not suggest any easy or predictable connection between these two senses of "employment": I do not argue that the job market will revive in English if every member of the MLA agrees to preface his or her book with the credo "I love literature, I really do," nor do I argue that the survival (or demise) of literary study depends on the utility of cultural studies to the professional-managerial class. But I do think literary and cultural studies can serve useful and even politically progressive ends, and I do think their ability to serve those ends is predicated in part on the economic and intellectual health of the profession. If the essays in this book prove moderately convincing, then, I hope that they will call readers to rethink how English—and English teachers—can best be employed.

Janet Lyon is my coauthor in chapter 8, but, of course, Janet Lyon is coauthor of everything I do; this time we're simply making it official. (Janet and I also need to thank audiences at the University of Chicago and the University of Virginia for their questions and suggestions in response to our tag-team argument in this chapter.) I owe many thanks also to Cary Nelson, Robert Dale Parker, Nina Baym, Claire Chantell, Lori Newcomb, Gerald Graff, Stephen Watt, Laura Kipnis, Jay Dobrutsky, Charles Harris, Curtis White, Michael Sprinker, Jeffrey Williams, Phyllis Franklin, Michael Vazquez, M. Mark, Eric Lott, Michael Denning, Cary Wolfe, Iris Smith, Ralph Cohen, Allen Carey-Webb, Don Hedrick, and Joe Tabbi for soliciting, reading, editing, improving, or just gainsaying various portions of this manuscript. I have relied time and again on the advice, support, and intelligence of a pair of good friends and trusted counselors, the truly cosmopolitan Amanda Anderson and Bruce Robbins. Niko Pfund and Eric Zinner have managed to exceed their reputation as terrific editors; Richard Powers and Mark Rykoff have been intercontinental pillars of strength—the one in Urbana, the other in Moscow. Thanks also to the Humanities Research Board of the University of Illinois at Urbana-Champaign for giving me the time to work on this book, and to the Cafe Kopi in downtown

Champaign for being such a wonderful place to work when you've got a whole day to yourself. Last but first, Nicholas and James, the two most astonishing and wonderful children I know, have kept me going and buoyed my spirits in innumerable ways I can neither tally nor repay. This book is dedicated to them, and to our Boys' Summer of '96, with all my love.

Portions of this book have been previously published. A section of chapter 1 previously appeared as "Aesthetics and the Literal Imagination," *Clio* 26.4 (1996): 439–53; chapter 2 appeared in *Social Text* 49 (1996): 75–95; a shorter version of chapter 3 appeared in the *Centennial Review* 40.2 (1996): 223–40. Chapter 4, with some modifications, appeared in the *minnesota review* 43–44 (1996): 131–44; a version of chapter 5, without the epilogue or introduction, appeared in *Critique: Studies in Contemporary Fiction* 37.3 (1996): 188–204. Chapter 7 is a revised version of an essay first published in *Advocacy in the Classroom: Problems and Possibilities,* ed. Patricia Meyer Spacks (New York: St. Martin's, 1996): 186–97. Chapter 9 appeared in *Transition* 69 (1996): 90–98, and chapter 10 was published on-line in the second issue of *ebr,* the electronic book review, at http://www.altx.com/ebr (1996). I am grateful to these journals, publishers, and websites for permission to reprint, refine, improve, and redact my work here.

Champaign, Illinois
April 1997

I

EMPLOYMENT IN ENGLISH

CULTURAL STUDIES AND CULTURAL CAPITAL

> The desire called Cultural Studies is perhaps best approached politi-
> cally and socially, as the project to constitute a "historic bloc,"
> rather than theoretically, as the floor plan for a new discipline.
> —Fredric Jameson, "On Cultural Studies"

In the past decade, cultural studies has named a desire, a desire Fredric
Jameson rightly links to the aspirations of populist intellectuals and
the utopian hopes of the Left. Yet cultural studies has also been presented
more "theoretically" (or prosaically) precisely as the floor plan for a new
discipline—a transdisciplinary or antidisciplinary discipline that promises
to remake the humanities and redraw or erase the traditional boundaries
between academic fields. What is arguably the most striking feature of cultural studies in the contemporary landscape, however, is the role it has
played in the collective disciplinary imaginary of literary studies: in the latter half of the 1990s, the project called cultural studies has come to name
not only a desire but also, and to the same extent, a pervasive fear.

The fear is a fear of dissolution, dissolution of the boundaries, the identity, the *quidditas* of literary study. After all, cultural studies, according to its own most common self-representations, has neither a methodology nor an object to call its own. It is quite possible, then, to understand the advent of cultural studies in literary studies as the amorphous outcome of three decades of intellectual debate in the field: as the discipline's notion of "text" expanded to cover a variety of materials formerly considered nonliterary or extraliterary, and as the discipline's methodologies increased in the number and variety of what once were called "extrinsic" approaches to literature, English departments have become places where a great variety of cultural texts are studied with a host of intellectual tools borrowed and modified from neighboring disciplines like history, philosophy, and anthropology. Its borders permeable on all sides, English has become an intellectual locus where people can study the text of *Sir Gawain and the Green Knight* from a Christian perspective, the text of the O. J. trial from a Foucauldian perspective, and the text of the Treaty of Versailles from a Marxist perspective. Appropriately enough, while cultural studies is hailed in some quarters as the means by which literary study can intervene in the social world of power, hegemony, and human affairs, cultural studies is decried in other quarters as the means by which a new generation of scholars will eventually eradicate whatever remains distinctive and "literary" about literary study.

In the pages that follow, I will not try to determine, once and for all, the correct formula by which English departments can blend the intrinsic with the extrinsic, the literary with the nonliterary, to greatest advantage. But I think it is a mistake to treat the prospect of cultural studies as a zero-sum game, as if scholars and students cannot spend their time and energy analyzing the social ramifications of a text unless they agree to neglect the text's formal and generic properties. The extent to which English departments incorporate the concerns of cultural studies will be the extent to which English departments institutionalize a mode of reading that asks after the production, reception, and social effectivity of texts; but the extent to which cultural studies becomes a mode of reading in literary study (as opposed to, say, a mode of reading in mass media and communications) will, conversely, be the extent to which cultural studies foregrounds the rhetorical operations of *literature*. It is entirely possible, in other words, to have your literature and your cultural studies too, if your literary study is cultural enough and your cultural studies is literary enough.

Those of my colleagues who fear that cultural studies will replace rather than enrich literary study, by contrast, are skeptical about precisely this point. As William Cain has recently charged, the arrival of new methodologies inevitably entails intellectual trade-offs, such that what we gain in the study of "culture" we must lose in the study of "literature":

> The Modern Language Association keeps insisting that the swerve toward cultural studies has not led to the displacement of this kind of close reading. But if you talk to teachers, it becomes evident that, in order to keep up with trends in cultural studies, they are cutting back on the time given to writers and books that students should be discovering and learning how to read. (B4)

Cain's formulation contains its conclusions, of course: on the one hand, we have writers and books students should learn to read, and on the other hand we have "trends" in cultural studies. To the first we must do justice; with the second we are merely "keeping up." This pretty much closes out the possibility that Stuart Hall, Eric Lott, or Janice Radway might qualify under *both* headings, as theorists in cultural studies whom students should discover and learn how to read. But as we shall see, Cain is far from alone in understanding the field in this way; and because our field is made up, in part, precisely *by* understandings of our field, we cannot chart the present and future of literary study unless we attend to why it is that cultural studies names both a desire and a fear.

Disciplines in the modern languages, in my view, should always be home to a variety of methodologies that ask *what* texts mean as well as *how* texts mean.[1] I am happiest, as a critic and as a reader, when I am learning how these two concerns are mutually illuminating—how the formal properties of a text are part of the work that text does in the world, and how its work in the world is enabled or conditioned by our understanding of its properties. But I am not narrowly prescriptive when it comes to asking what kind of work *English departments themselves* might do in the world. I believe there are any number of ways to introduce students to the demands and delights of close textual study, and it is of little concern to me whether our students start by reading Wordsworth and work their way to deconstructions of contemporary representations of race, or start by analyzing Madonna videos and work their way to an understanding of the Romantic crisis lyric. Accordingly, I do not lose much sleep worrying about whether my students (graduate or undergraduate) will carry on the work of literary study in the way I like most to see that work done. Nonetheless, it is clear to me that our disciplinary desires and fears are driven as much by our projections of the future as by our assessments of the present. The controversy over cultural studies is thus part of a more general crisis of reproduction in the modern languages—a crisis whose occasion is the question of whether there is any useful social purpose served *either* by literary study, narrowly conceived, or by cultural studies, broadly conceived.

A crisis of reproduction? Doesn't that sound awfully melodramatic? Perhaps things are at bottom much simpler than that; perhaps it's merely that many of our field's major theorists, from Frank Lentricchia to Wendy Steiner to Edward Said, have rightly dissociated themselves from the excesses of "politicized" literary study and turned our attention once again to art, to beauty, to the purposive purposelessness of the play of forms. Is there really any reason to call this the occasion of yet another "crisis," particularly in a field that *always* thinks of itself in terms of crises?

I believe there is; I think there's more going on here than just a return to art, and I think we can begin to understand what it is if we attend to a certain generational anxiety that defines contemporary fears of cultural studies. The figure to which I want to call attention (and on whom I hope to keep your attention for the remainder of this book) is the figure of the graduate student: here, the figure of the graduate student who no longer knows—or, worse, no longer desires to know—what might be "literary" about literary study.

William Cain's account of the field depends heavily on just this figure:

> Part of the problem is that the graduate students who become our faculty members are not prepared to teach close reading.

They have not learned the skills as undergraduates and, unfortunately, no one in graduate school has encouraged them to make up for their lack. I scan hundreds of transcripts when we make faculty appointments, and they reveal a numbing non-literary sameness—a compilation of graduate literature courses that are really courses in sociology, media, postcolonial politics, and the like. Courses on sexuality are everywhere. But I rarely detect courses on the literary subjects that graduate students might eventually teach in classes of their own. (B4)

And his conclusion is as sweeping as it is stark: "When a graduate student leaves the university with a Ph.D., he or she has little idea of what it means to read a text carefully or how to convey to students the skills needed to perform this activity. Nor is he or she prepared to make the choices required when designing courses and curricula for undergraduates" (B4–B5).

It is tempting to surmise that Cain's department of English at Wellesley College must have done some truly unfortunate hiring in the past few years, but similar reports up and down both coasts convince me that the phenomenon is not confined to Wellesley. From one prestigious, public eastern university comes the report that graduate students now study queer theory more than any other department "specialization"; from another prestigious, public western university comes the report of a department riven between people who want to jettison literature from the curriculum and people who want to jettison everything *but* literature. And from the University of Washington comes the following report from Ross Posnock:

My work and teaching blend literary criticism and intellectual history in an English department where the ideology of cultural studies, as described by Jameson, clearly has enthralled the majority of graduate students. In English departments the embarrassed, defensive status of the intellectual is matched by the low repute of literature (indeed of the aesthetic itself) and of those who dare construe their job as primarily devoted to its internal explication and external contextualization.

In the postmodern regime of English studies the intellectual, literature, the aesthetic, intellectual history are all held under suspicion on grounds of complicity with the enemy, which includes various instruments of white male power—universalism, cosmopolitanism, elitism. (18–19)

It would seem, then, that the students invoked by Posnock are precisely the graduate students who apply for tenure-track jobs at Wellesley: disdaining literature and those who teach it, they nonetheless seek jobs in departments of English precisely in order to liberate English from literature, and to offer ill-conceived and poorly designed courses on cross-dressing, Chicana/o graffiti, DisneyWorld, and the politics of postcolonial poststructuralism.

I do not deny that literary studies are in need of defense, and I do not deny that there are some departments of English that house and foster all the sins Posnock and Cain enumerate. But I do want to look more closely at what these defenses of literary study are in fact defending; as I will show when I discuss the introduction to George Levine's important edited collection, *Aesthetics and Ideology,* there remains a profound ambivalence, even among defenders of the literary and the aesthetic, as to whether "the aesthetic" is important because of the uses it serves in legitimating a domain for literary criticism, or because it serves no use whatsoever. Before I discuss Levine's introduction to *Aesthetics and Ideology,* though, I'd first like to file a contrary local report on the state of the discipline, just for the record.

At the University of Illinois at Urbana-Champaign, we have done a fair amount of hiring after a few years of severe budgetary constraints; we have hired Ph.D.s from a wide variety of institutions, in a wide variety of fields from medieval literature to queer theory. I hope I will not flatter my junior colleagues unduly if I insist, *contra* Cain, that every single one of our recent hires knows how to read a text carefully, and almost every single one of those hires has been extremely successful in the classroom; every semester, the Illinois student newspaper publishes the names of faculty and graduate students who receive particularly strong course evaluations, and since 1990 the list from English has included almost every junior faculty member we've hired. They must be doing *something* right, surely—and because our junior faculty are reviewed by their senior peers every year, I happen to know that they're quite capable of designing courses and curricula on their own. I mention this not to brag about our good fortune, however; I mention it because even though our department has fared very well in hiring smart theorists who are also good teachers, some of my senior colleagues *nevertheless* perceive a disjunction between theory and pedagogy, and worry accordingly that the recent "drift" of the department has not been good for undergraduate education. If William Cain's department is haunted by fears that it is no longer possible to hire a Ph.D. in English who's a good reader, then, in a much milder manner, so is mine—even though we have no younger faculty who would actually justify this fear.

In my department, in other words, this fear cannot be gauged by measuring the level of happiness or discontent with regard to actually existing junior faculty; it is too nebulous to be focused on any individual person—until we come across that one job candidate who reads poetry for the "cultural text" but doesn't know much about prosody, that one post-something theorist whose campus-visit presentation was difficult to understand. *Then* the discussion begins, and people wonder what other Ph.D. programs must be *thinking* these days, and *how* will our undergraduates ever be able to learn from these incomprehensible young turks, and what will become of literary study once we titans no longer roam the earth . . . And after a while *I* begin to wonder, how long have people harbored these fears, waiting for them to find an object? For ten years the department hires one good undergraduate teacher after another, and most of them compose syllabi full of works of literature (as opposed to, say, videos of the O. J. trial), which they train undergraduates to read closely, and *now* we have an anti-theory backlash? Apparently a very few of my colleagues had been waiting a long time to vent their fears about the horrible things that are happening to the profession, but hadn't yet found the chance.

Usually these fears circulate around hiring and tenure, and they are tied as well to the question of whether new hires, in a research university, should be driven by the needs of the undergraduate curriculum (a Miltonist hired for a Miltonist retired) or by the research developments in theory and criticism (a queer theorist hired for an Augustan scholar

9

retired). But on one occasion in the spring of 1994, when the department was charged with rewriting its bylaws (a job I regarded, at the time, as the intellectual equivalent of cleaning out the basement), we suddenly found ourselves in the midst of a substantive discussion over the content of our self-description: was it fair, we wondered, to describe ourselves as offering instruction in English and American literature, or should we say "literature in English," even though we offer so few courses outside the Anglo-American spectrum? Should we say "literature and criticism" or "literature, criticism, and interpretive theory"? How should we describe our offerings in film? And last but first, should we amend "literary studies" to "literary and cultural studies," and if so, how should we recognize cultural studies in the curriculum?

These questions become all the more urgent when we turn to graduate study, where, indeed, Illinois has seen a good deal of variety in recent dissertations, and a great deal of speculation about the relation between dissertation topics and jobs. As I'll explain in more detail in chapter 4, the job market is such that graduate students feel compelled to write extremely specialized dissertations even as they will likely be asked, if and when they get a job, to teach fairly general, unspecialized courses. But even among our graduate students whose work is most specialized and/ or most inflected by critical theory, Illinois dissertations have been (like the popular T-shirts) largely literary. One of our most talented students did write a dissertation on British music halls and the emergent dis-courses of professionalism and propriety at the turn of the century, and got a job only after a number of frustrating years of searching; two of our other students whose work might fall under the cultural studies heading wound up writing on contemporary gay and lesbian literature, in one case, and contemporary anthologies of erotica marketed to women of various ethnic groups, in the other case. The vast majority of the rest of our students have been writing more or less traditional dissertations on novels, poetry, and drama of various periods; some are influenced by new historicism, some by feminism, some by Marxism, some by recep-tion theory. None, so far as I know, are inclined to suspect literature of complicity with the enemy. And as for our new departmental self-description, it now reads like this: "The Department of English is organized to provide instruction in literatures in English, literary theory and criticism, the English language, expository and creative writing, writing studies, English Education, film, cultural studies, and closely related fields." Those last four words represent all the minor compromises left over after the department had hashed out (in many committee sessions and then in a full faculty meeting) the relative place of linguistics, teacher training, theory, creative writing, business and technical writing, film, and (oh yes) literary and cultural studies; but the most controversial items, which not coincidentally are the focus of my attention here, were "literatures in English" and "cultural studies."

So much for my contrary local report. For what it's worth, it may serve as evidence that many departments of English may be troubled in one way or another, but are not quite as absurd as Cain or Posnock suggest. And yet it cannot be denied that the discipline has been indelibly changed by the past ten years alone, ever since deconstruction moved from the avant-garde of the field to the lingua franca of the culture, ever since Foucault and Gramsci (via new historicism, queer theory, and cultural studies) became the major discursive options for theoretically inflected cultural analysis. The discipline's critics are not entirely wrong to suggest that in the present regime, one's theoretical allegiances can determine one's critical conclusions: either you believe in the forces of containment and recuperation, in which case it becomes your job to show how the seemingly "liberatory" or "progressive" aspects of the culture ultimately serve the conservative purpose of perpetuating a political order in which "freedom" is but a name for a particularly deceptive form of self-policing, *or* you believe in hegemony and resistance, in which case it becomes your job to show how the seemingly "repressive" or "reactionary" aspects of the culture ultimately can be made to serve surprisingly (yet reassuringly) liberatory or progressive ends.

In this dispensation it should come as no surprise that literary texts are commonly treated as pieces of cultural evidence rather than as artifacts to be explicated on their own terms (however their "own terms" may be construed). In and of itself, there is nothing wrong with treating literary texts in this way: they undoubtedly *are,* among other things, important pieces of evidence about the culture(s) from and to which they speak, and any reasonable historicist, feminist, reader-response, or psychoanalytic critic will say so. (Even myth critics, if there are any left on the planet, will agree.) On the other hand, there may indeed be something wrong with forms of cultural analysis that seem to dictate their conclusions in advance of their evidence, and there may indeed be something wrong with analytical procedures that fail to attend to the specific details of what *kind* of evidence is placed on the table. It may be folly to claim that English departments are places where graduate students hold literature under suspicion, and where jejune junior faculty are incapable of constructing a literature syllabus. Nevertheless, it is possible to ask a skeptical question about English in a different register: is the discipline dominated by reading practices that so determinedly overlook the specificity of textual genres (be they novels, verse satires, Hollywood screwball comedies, sonnets, epistles, mystery plays, manifestos, billboards, or laws) that, for the purposes of those reading practices, professors of English could just as well be reading anything at all?

George Levine's edited collection *Aesthetics and Ideology* speaks directly to this question; one might even say that the book is itself, like the formation of the Association of Literary Critics and Scholars in 1994, a major announcement of a major scholar's dissatisfaction with current practices in literary studies. But Levine himself is no nostalgic belletrist, and he has little sympathy with the ALCS theory that English departments have been taken over by rabid ideologues who hate great literature. How then, he asks in his introduction to the volume, "Reclaiming the Aesthetic," can he call his colleagues to a renewed examination of aesthetics without being mistaken for a reactionary and an enemy of the people? How might he best frame his misgivings about contemporary criticism while retaining his political allegiances to feminism, multiculturalism, deconstruction, new historicism, gay and lesbian studies, and all their friends? "This book," writes Levine,

> and this introduction, have required that I face directly my own anxieties about what my passion for literature will seem like to the critical culture with which I want to claim alliance....

Beginning this book with the language of the affective, the sublime, the aesthetic, I hoped to rescue from the wreckage of the mystified ideal of the beautiful the qualities that allowed for such rich ambivalences. Eliot is anti-Semitic and worse. Arnold is both statist and snob. I wouldn't be without the writings of either of them. That, I recognize, puts me and this book under suspicion. (11)

Levine occasionally gets quite dramatic on this score: the "anxieties" Levine faces seem to derive principally from the fear of losing face before one's valued colleagues, and one would think, from reading Levine, that one's valued colleagues are only all too ready to pounce on whoever starts talking about anapests and pastoral elegies instead of gender and hegemony. As Levine writes toward the close of his essay,

I am happy to see politics as an inescapable element of all human creation and to read every text into its political moment. But I ask, breathlessly and nervously, for the opportunity not, as I try to come to terms with the specific forms of literature, to use my understandings of these texts in a political program that turns them into instruments and destroys that very small breathing space of free play and disinterest left to those who risk finding value even in the literature that seems to despise them. (21)

Breathless and nervous, Levine appears to be under even heavier assault than are Ross Posnock and William Cain. All you have to do in an American English department, it appears, is to profess aloud your abiding love of literature, and the room will fall strangely silent; within days you will be labeled a revanchist; by the end of the term you'll have been booted out of the fancy critical clubs, and come December you can just forget about getting that table by the window at the MLA's annual Banquet of Critical Eminence.

There are, of course, a few serious questions buried in my somewhat facetious response to Levine's somewhat theatrical "staging" of his book: does the politicization of criticism *require* a devaluation of aesthetics? Is it necessary to overlook the specific properties of literature in order to read literary works in terms of their relations to larger cultural formations?

CULTURAL STUDIES AND CULTURAL CAPITAL

Any assessment of the profession—any assessment of the functions of criticism at the present time—will turn on how those questions are answered. Roughly half the profession's accusers seem willing to indict any and all "political" criticism, on the grounds that "politics" is precisely that which is bracketed or transcended by the monuments of timeless aesthetic excellence. The other half of the profession's accusers make a more careful case, in which the politicization of literary study is a problem of degree rather than of kind: literature and criticism are inevitably entangled in social, historical, and ideological commitments, but contemporary literary criticism simply stresses this aspect of literature too strongly, just as an earlier generation of critics failed to stress it strongly enough. The first of these positions, in my view, is either contentless or intellectually bankrupt, and not even the ideologues of the Right, who so often use the aesthetic as a stick with which to beat "politically correct" literary criticism, believe a word of it. The second position seems to me quite plausible. And it is this position, apparently, to which Levine wants to claim allegiance as he tries to straddle aestheticism and historicism, formalism and feminism, the pleasure of the text and the epistemology of the closet. Unfortunately, his discussion of the role of aesthetics serves only to blur the distinction between position one and position two, and thus to invite, from politically minded theorists, the very kind of dismissal or condemnation it most fears.

By my count, there are at least half a dozen formulations of "aesthetics" in Levine's introduction, and half a dozen corresponding diagnoses of what's wrong with the profession. This in itself is not a bad thing, but it does indicate how difficult it is to devise a firm disciplinary basis for the protean thing called literary criticism. In his very first paragraph, for instance, Levine remarks that he "conceived this book in response to the radical transformation of literary study that has taken place over the last decade," in part because that transformation marks "a change that might challenge the very existence of departments of literature in universities." The paragraph closes with three questions that flow from this concern: "Can, in fact, a category, literature, be meaningfully constituted? If so, once constituted, is it worth much attention? Is not, after all, the real subject of literary study ideology, the real purpose political transformation?" (1). The first two of these are very good questions indeed; as I hope to have made clear thus far, I would like to inhabit a profession that asks them in the most rigorous possible manner. But the third question is a strict non sequitur; it even contains a non sequitur within a non sequitur. For if in fact it is not possible to constitute literature as a distinct category of writing (and many minds finer than mine have tried to do just this), there is no reason to assume that the subject of literary study is therefore ideology (whatever "ideology" might mean in this context); and even if the subject of literary study *were* ideology, there would be no warrant for eliding the study of ideology with the goal of political transformation. Many professors in history and political science, I believe, manage to study ideology without demanding to see their students' voting records at the end of the semester. In the course of only a few lines, then, Levine has moved from asking fundamental and indispensable questions about the *donnée* of the discipline to caricaturing politically committed criticism in terms as reductive and (mis)leading as *New Criterion* boilerplate.

Levine then moves to more specific targets, naming Fredric Jameson, Edward Said, Stephen Greenblatt, and Eve Kosofsky Sedgwick as examples of politically committed readers. But as Levine immediately adds, his quarrel is not with these people, whom he credits with having "wonderfully enriched the possibilities of literary criticism"; his quarrel is with "their followers" who "reduce critical practice to exercises in political posturing" (2). Here again is that distinctive generational anxiety, the impulse to scorn the sort now growing up, all out of shape from toe to top: their parents had some redeeming qualities, but evidently those qualities were recessive, for the children display only the vices of their elders in exaggerated form. Yet three pages later, it turns out that Levine does have a quarrel with Sedgwick after all. Calling her work "a rich and illuminating criticism that makes literature more, not less interesting," he nonetheless registers a complaint about its purpose: "like much of the best criticism today, it is, however, using literature primarily as a means to broad cultural conclusions" (5). This is so vaguely worded as to be self-defeating: it encompasses not only Sedgwick but Kermode, Frye, Trilling, Auerbach, Leavis, Eliot, and Arnold, all of whom saw "broad cultural conclusions" as a crucial element not only of literary criticism but of literature itself. *Mimesis,* after all, is a magnificently capacious term. So unless one wants to indict critics for drawing broad cultural conclusions from Blake, Stevens, Lawrence, or the metaphysicals, one will have to rewrite the indictment and submit it anew.

As it happens, that's pretty much what Levine does in the remainder of his introduction. First, he moves back onto firmer ground, asking about the value of literary texts as evidence for cultural conclusions: "my remarks are not to question Sedgwick's analysis of homosocial desire, but to require attention to how the overall argument of *[Between Men]* . . . implies very complicated arguments about the way literature works in culture" (6). In other words, if a good close reading can tell us a lot about a text's rhetorical and ideological operations, what might we legitimately infer about how those operations might have refracted, resisted, or affected the culture(s) in which the text was written and read? "The question of what sort of evidence about culture 'literary' texts provide has by no means been resolved," writes Levine (6–7). What, then, is the difference—if there is one, or only one—between reading treaties and treatises, and reading Defoe's *Robinson Crusoe,* as texts concerned with the English slave trade?

Had Levine stopped here, his introduction would have set us the worthy task of determining the cultural and evidentiary status of literature—a task sometimes neglected by cultural studies, partly because cultural studies too often attempts an analysis of "reception" without an explicit theory of reception aesthetics (the road not taken in American criticism thus far) or a theory of genre (a road neglected since Frye, with two major exceptions—Fredric Jameson's neglected revision of Frye in *The Political Unconscious,* and the dissimilarly neglected work of Ralph Cohen). But Levine does not stop here; instead, he goes on to ask whether it's worth reading literature at all "if what we want to know can be discovered through other materials," and then to ask, apropos of D. A. Miller's *The Novel and the Police,* why we should bother reading literature if we find it complicit with things we don't like:

Miller's answer, as I understand it, is that [the Victorian novel] enables readers to see how this bourgeoisifying project is carried on in other, contemporary, forms. In effect, the point of reading literature sensitively is to warn readers against reading literature. If one objects to the idea of the "liberal subject" and to the political regimes that rely on it, why further propagate texts enlisted on the side of the enemy? (9)

This is another set of questions altogether. It's one thing to ask whether and how literary texts can provide us the material for "broad cultural conclusions"; it's quite another thing to ask literature to provide us with a unique kind of informational content that can be found nowhere else in the world. And then it's another thing to ask whether it makes sense to read influential texts and authors whose influence one may regret or wish to contest. And then it's yet *another* thing to ask, as Levine does two pages later, whether the "aesthetic" might be a discursive realm of relative autonomy from purposiveness:

Does literature have any standing that might, even for a moment, exempt it from the practical and political critiques to which all other artifacts of culture are apparently subject? When art seems, directly or indirectly, intentionally or inadvertently, openly or surreptitiously, to sustain, create, justify, or forward politically or socially objectionable ends (from whose perspective?), are there any grounds for giving it the privilege disallowed to other enemies of the good, the true, and the just? (11)

Here too, there's a trenchant question buried in an infelicitous formulation: what does Levine mean by "exempt," and what does it mean to be exempt *for a moment?* Surely no working critic, not even the most besotted follower of Jameson, Said, et al., is so foolish as to call the police when Macbeth murders Duncan. So literature is probably exempt from the demands of quotidian practicality in that sense. But what kind of momentary exemption should we grant Ezra Pound when he writes, in his *Pisan Cantos,* that the goyim are sheep and are led to slaughter in great numbers by international Jewry? Are we to forswear our "practical and political" responses to this kind of profound obscenity, as did the

1948 Bollingen Committee, on the grounds that Pound's work, as poetry, should be considered "exempt" from such concerns?

At this point we're not yet halfway through Levine's introduction, and perhaps we will be forgiven if we confess to confusion: what exactly is wrong with contemporary criticism? Let's review what we have thus far. Does it neglect to ask whether literature is a meaningful category? Does it attempt political indoctrination as a result? Is it too cavalier about deriving broad cultural conclusions from relatively little documentary evidence? Is it inattentive to the possibility that its informational content might be unique? Does it attend to major writers merely to unmask their complicity in evil, and is it heedless of the dangers of further propagating the enemy's texts? Or, having failed to ask the first of these questions, is criticism blind to the possibility that literature might indeed have a "special" status that exempts it from "practical and political critiques"? These questions, I should note, are not mutually exclusive: one can certainly say all these things about contemporary literary criticism, and one can even add, if one has a mind, that it's poorly written, carelessly footnoted, inelegantly punctuated, shoddily bound, callously marketed, shamefully reviewed, and brutally expensive, too. But not all of these objections—Levine's real ones, or my petty hypothetical ones—constitute or even license a close examination of the aesthetic.

Our confusion is real; it is, in many ways, the occasion for this book. Why, indeed, should the aesthetic be a critical component of the disciplinary definition of literature, as distinct from music, dance, or the plastic arts, where form is easier to distinguish from propositionality? More curiously, why should an aesthetic definition of literature be a critical component of the franchise of the English department? After all, we do not live in a world where university trustees, legislators, parents, and journalists rise up in arms whenever English professors fail to do justice to the sublime and the beautiful; as I'll point out again in chapter 3, public furor over the mission of English rarely addresses anything other than basic writing courses—with the occasional exception of your standard-issue right-wing media campaign that alerts Americans to the shocking fact that Shakespeare is no longer taught in literature classes. What, finally, is the relation between the subject matter of literary-slash-cultural studies and the public legitimation of the discipline known as English? What is the relation between the field's internal self-definition and its external constituencies?

We (I) cannot answer these questions directly, because they admit of no definitive answer; but we (I) can describe their parameters and suggest what's at stake in trying to grapple with them. Fortunately, the reason the questions are not directly answerable is intimately tied to those parameters: the potential constituencies of the field depend largely on what the discipline of English means *institutionally,* as a subject in college and high school courses, and what English means institutionally is dependent in turn on a congeries of social and economic movements well beyond the control of any one professor, department, or syllabus.

To grasp the relation between the subject matter of the field and its public legitimation, then, we need to inquire into the status of literature as cultural capital, as John Guillory has done in his landmark *Cultural Capital: The Problem of Literary Canon Formation.* Guillory says almost nothing about cultural studies *in nomine,* but his book does theorize the relative "decline" of literary studies so provocatively as to afford us an explanation of why cultural studies might have become, for much of the discipline, the ideal self-description *du jour.* For Guillory, the status of literary study is inseparable from the larger social conditions that make literary study either valuable or superfluous as cultural capital:

> It has proven to be much easier to quarrel about the content of the curriculum than to confront the implications of a fully emergent professional-managerial class which no longer requires the old cultural capital of the bourgeoisie. The decline of the humanities was never the result of newer noncanonical courses or texts, but of a large-scale "capital flight" in the domain of culture. . . . The professional-managerial class has made the correct assessment that, so far as its future profit is concerned, the reading of great works is not worth the investment of very much time or money. The perceived devaluation of the humanities curriculum is in reality a decline in its *market* value. If the liberal arts curriculum still survives as the preferred course of study in some elite institutions, this fact has everything to do with the class constituency of those institutions. (45–46)

Let me flesh out Guillory's analysis with a brief anecdote, mediated by way of Richard Ohmann's observation, in *Politics of Letters,* that

> for Wesleyan students (and for those at Yale, Stanford, Wellesley, etc.) there is *still* no penalty for pursuing the humane and pleasant activity of reading good books and trying to understand the world. These students have a reserved place waiting for them in the professional-managerial class or the ruling class, some by virtue of having made it into an elite college, most by birth and nurture. (12)

When I first went "on the market" at the 1988 MLA convention in New Orleans, I interviewed for jobs at a wide array of schools, and nothing made the stratification of the profession so palpable to me as the answers I got when I asked my interviewers about the number of their English majors relative to the total number of undergraduates at their institutions. On one fine December morning I managed to go from an interview at Williams College, with more than 250 English majors among 1500 upper-division students, to Auburn University in Alabama, with an undergraduate enrollment of 21,000 and 120 English majors. Those numbers alone, it should be noted, determine much of the working conditions of faculty in English at both institutions: faculty at Williams are invited and expected to teach in the area of their "specialization," and though their school generally values their teaching more highly than their research, the number of English majors enables the college to institutionalize a diverse array of advanced courses in English, whether these be courses in Restoration drama, film noir, or postcolonial theory. At Auburn, by contrast, the range of advanced courses is limited not by the research interests of the faculty but by the cultural capital of English at Auburn, and faculty accordingly teach more courses than their counterparts at Williams—and *many* more introductory courses, including courses in basic writing. Even at individual institutions, then, the content of the curriculum is determined largely by the status of English as cultural capital—or, more accurately, cultural capital is realized and invested *as* cultural capital precisely by means of individual institutions operating dynamically within larger institutions.

On one hand, the implications of this point are trivial, and everyone in English knows about them—just as everyone knows that the teaching load at Williams differs from the teaching load at Auburn. One might say, for instance, that Cain, Posnock, and Levine are worried about something that can be an issue only at relatively "elite" institutions, where cultural studies appears as a curricular option unavailable to colleges whose English curriculum is weighted heavily toward introductory courses, and where the question is not "how can we get our students to stop reverencing literature and start paying attention to the social text," but "how can we get our students to pay attention to literature in the first place?" But on the other hand the point is fundamental to the role of cultural studies in English, insofar as cultural studies does *not* have to be confined to elite institutions, and can be as central to an introductory curriculum as to an advanced course of study; similarly, the point is fundamental to the constitution of English departments in the United States, insofar as the franchise of English depends on the *institutional* capital of English in specific institutional locations.

I will return to these issues in the chapters that follow, as I turn to the employment of English in specific institutional locations. For now, though, I want briefly to address the status of cultural studies as cultural capital. Guillory's analysis takes for granted one of the premises of the New Right as articulated most clearly in William Bennett's *To Reclaim a Legacy,* namely, that the humanities are in decline. Guillory rightly argues that this alleged decline of the humanities has nothing to do with the introduction of noncanonical works to the literature syllabus, and everything to do with the cultural capital of literary study and its relation to productive capital (that is, money) for college students. Hence Guillory's attribution to the aspiring professional-managerial class of the sense that "the reading of great works is not worth the investment of very much time or money." But what happens if we contest the narrative of decline at the outset, and try to account for the resurgence of undergraduate interest in the English major in the 1980s and 1990s?[2] Surely it would be tempting but wrong (in Guillory's terms and in mine) to attribute that resurgence solely or chiefly to the newer multicultural curricula in English.[3] Perhaps instead we might point to two general

21

economic factors that may have swelled enrollments in English in the past ten years—first, the widespread (but ultimately mistaken) belief that there would be a "faculty shortage" in the 1990s, such that employment in colleges (of great concern for graduate study) and secondary schools (of great concern for undergraduate study) would be a likely prospect even for people who received degrees in the humanities; and second, the widespread (and ultimately well founded) belief that the global economy was producing jobs that were less stable, less secure than the jobs of forty years ago, such that for some areas of nonacademic employment, a general liberal arts degree might be seen by prospective employers as more attractive than a degree that signified a college career of technical-vocational training.

Let me add to these observations the following questions: who, exactly, was "credentialed" by universities back in the days when the humanities were not in crisis? For whom was literary study a form of cultural capital? Might it not be the case, as Francis Oakley has suggested, that the rise in professional-vocational courses of training since 1970 (and the relative "decline" of the humanities) coincides with the arrival at universities of vastly more diverse student populations (particularly more diverse with regard to class origins) beginning in the late 1960s? My hope in raising these questions is not to claim that we're just fine in the humanities these days, thank you, despite everything you've heard to the contrary. Rather, my hope is to raise questions about Guillory's account of the relation between literary studies and cultural capital just as Guillory has raised questions about the relation between cultural capital and the canon. If it is true, as Guillory claims (as I would claim as well), that the status of literary studies as cultural capital does not depend, solely or chiefly, on the curricular content of literary studies, it may also be true that there is some degree of independence between the status of literary studies as cultural capital and the employability of a degree in English. It is possible, I am claiming, that "literature" may indeed have declined in cultural authority but "English" remains a potentially valuable career asset. To put this in more colloquial terms: whatever the status of "literature" as an index of cultivation and class status, degrees in English may still be convertible into gainful employment—

not because they mark their recipients as literate, well-rounded young men and women who can allude to Shakespeare in business memos, but because they mark their recipients as people who can potentially negotiate a wide range of intellectual tasks and handle (in various ways) disparate kinds of "textual" material, from memos, legal briefs, and white papers to ad campaigns, databases, and electronic newsmagazines.

And if we want to gauge the relative status (as cultural capital) of literature and cultural studies, we should have yet one more question for Guillory's account of the field. If, as *Cultural Capital* claims, the new professional-managerial class no longer requires the old cultural capital of the bourgeoisie, then it is not clear whether *all* kinds of cultural capital are now utterly superfluous to the accumulation and distribution of productive capital (as Bill Readings emphatically argues in *The University in Ruins*), or, by contrast, whether a redesigned curriculum in the humanities might actually be of greater use to the credentialization of the professional-managerial class. In forwarding the latter suggestion (since I disagree strongly with the former),[4] I do not want to be understood as saying anything so simpleminded as "we must substitute Toni Morrison's *Beloved* for Milton's *Lycidas* because this is what the new global economy requires"; to date I have heard of only one employer who asks such things of his job applicants: Milton Rosenberg of Chicago, who quizzes prospective college interns for his radio program as to whether they can name three nineteenth-century British novelists. (When I asked him whether it might not be more pertinent to a job in mass media that students be able to name three contemporary non-American novelists writing in English, I was dismissed in somewhat predictable terms, but I do not generalize from this that William Thackeray and Elizabeth Glaspell are necessarily more conducive to mass media employment than Margaret Atwood and Chinua Achebe.) Rather, I want to ask whether the advent of cultural studies can be understood as a response to the market value of literary study. This is not simply a question of whether English professors are offering courses on music video in order to remain somehow "relevant" to the cultural lives of their students (though it does participate in that question to some extent); more fundamentally, it is a question of whether the distribution of cultural capital serves a purely

CULTURAL STUDIES AND CULTURAL CAPITAL

discriminating function, to naturalize and legitimate socioeconomic inequality, or whether the content of cultural capital might matter in some substantive way to the traditional liberal-progressive project of "critical thinking."

Bruce Robbins raises precisely this question in his trenchant review of Guillory's book: does the content of the curriculum *signify* in a meaningful sociopolitical sense, or is the curriculum primarily a means of marking and enforcing distinctions among our students regardless of whether those distinctions are built on Shakespeare or on Madonna? Robbins writes,

> [E]ven if we believe that knowledge of Latin and Greek was no more than an empty diacritical mark differentiating rulers from ruled, it does not follow that either vernacular literature over the past seventy-five years or the content of today's far more democratically accessible curriculum is equally irrelevant, that these too could serve only to differentiate. And if they are indeed functional, then it would also seem, for example, that the ability to get one's own experience re-classified as part of cultural capital—which is one description of what multiculturalism is about—should also be classified as a genuine if not necessarily momentous redistribution of power. From the moment when knowledge of rap music or rape statistics or the genealogy of the word "homosexual" is measured on examinations and counts toward a degree, there has been some change, *pace* Bourdieu, in access to credentials. (373)

Note that neither Robbins nor Guillory is making the obvious (and, by now, quite tired) argument about curriculum change—the demographic argument, which says that new student populations require new course offerings. Both are rightly skeptical of that "representational" logic. Instead, they ask, as I am asking, whether the content of the curriculum in the humanities might have any important relation to the social standing of the humanities (and, as a result, to the health of the humanities in terms of enrollments, job placement, faculty lines, and cultural authority). Can we say that on some level, the profession somehow *knows,* structurally, that its prestige is not what it once was (for whatever reasons), and is trying to recoup some of its lost authority by redefining its object—not as the study of great works, but as the enhancement of rhetorical techniques of interpretation that can be applied to a vast variety of cultural "texts"? David Simpson, in his recent book *The Academic Postmodern and the Rule of Literature,* argues that this is exactly what's going on when English departments start including "cultural studies" in their self-descriptions:

> Both in Britain and the United States, humanities intellectuals are particularly vulnerable, because of our association with those very qualities [Thorstein] Veblen attributed to us, qualities of the occult and the liminal. This shared anxiety must be part of any comprehensive explanation of our current turn to cultural criticism, which paradoxically provides us with a rhetoric of referentiality, a posture of speaking about the world, even as we admit that the world is made up largely of representations. Cultural studies, as we now see it, is a form of survivalism, and those who deplore its incursion into the universities would do well to reflect upon the degree to which it ensures their own continued existence. (7)

Cultural studies as survivalism: that, in a nutshell, is both the desire and the fear driving the debates about cultural studies, as far as the near future of English is concerned. In this corner, David Simpson, cautioning us to think that cultural studies might actually be propping us up; in the far corner, George Levine, cautioning us to think that the prop might just deprive us of our public legitimacy as stewards of the aesthetic.

Interestingly, this polarized scenario comes in other versions; more interestingly still, they all lead to the same dystopian conclusion, as I'll show in a moment. From the Left, especially the British Left most closely identified with cultural studies, there is the well-founded fear that in American universities, "cultural studies" will eventually come to be understood as a rough synonym for "the humanities." The fact that the term has been appropriated by English departments, anthropology departments, communications and speech departments, and the foreign languages (where it is often seen as very much a survival mechanism that might compensate for the drop in enrollments for advanced language study) is ample evidence, to this wing of the academic Left, that the term "cultural studies" has lost its critical force and has become another name for business as usual on the leafy side of the quad. Conservatives have a similar story to tell. From Alvin Kernan's book *The Death of Literature* to the proclamations of the ALCS, the message is pretty much the same: the kids just aren't reading anymore. They come to college, their heads brimming with Beavis and Butt-Head but without the faintest notion who Chaucer is; and the professors, deserting their posts as educators, have given up trying to offer students the great books of the ages, either because they no longer believe in the concept of great books, or (more cynically) because they no longer believe they can sustain enrollments by professing the faith of the great books. Accordingly, departments of English are gradually becoming places that offer courses on Barbie dolls, rock stars, and the Disney empire, while seekers of truth pine desperately for the courses in Milton and Wordsworth the department no longer offers.

At least in this one respect, both the left and right wings of the culture wars can look at the American academy, hold their heads in their hands, and bewail what cultural studies hath wrought.

The reason Guillory's theory of cultural capital is so germane to the future of literary study, then, is not that it clarifies matters but that it helps to explain why matters are as cloudy as they are. Depending on how one draws the connections among literary study, cultural capital, and meaningful employment, one can lament the turn to cultural studies and call for the restoration of literature to the central place in the curriculum, on the grounds that this will at once preserve the intellectual legitimacy of English and revive the discipline's level of public support, *or* one can applaud the turn to cultural studies as that which will preserve the intellectual legitimacy of English and revive the discipline's level of public support. And, in turn, each of these options can be cast quite differently, as a blueprint not for the revival of English but for its demise.

In *The Western Canon,* for instance, Harold Bloom has mused that departments of English will gradually be pruned of their faculty who love literature until they are composed of small handfuls of close readers, whereupon they will have roughly the size and influence of departments of classics—which they will resemble in many other antiquarian respects. At the same time, critics of cultural studies who combine George Levine's sense of the franchise of English with a healthy distrust of contemporary cost-cutting measures in academe claim that if literary study eventually becomes synonymous with cultural study, the department of English will be amalgamated with any number of cognate departments in the humanities, from anthropology and film to communications, rhetoric, and media studies. And then along will come some enterprising dean or provost who will look at all this "duplication," all this mess of faculty purportedly studying the same thing called "culture," and we will find our numbers cut in half as the university moves to eliminate or reduce some of its nonrevenue-producing activities like offering degrees in the humanities. Yet alternatively, if literary study follows the Bloom model, remaining *literary* study and refusing to cave in to the blandishments and temptations of cultural studies, departments of literature will find themselves becoming small, ineffectual side offerings in the general university market. Their faculty will be few and powerless, but proud and noble.

How can it be that both scenarios end in the destruction or evisceration of literary study and the humanities? If English departments embrace cultural studies in an attempt to appeal to the cultural needs of the professional-managerial class, they may be seen as duplicating the knowledges produced in other departments, and downsized or eliminated accordingly. And if English departments abjure cultural studies, they may be seen as irrelevant to the cultural needs of the professional-managerial class, and downsized or eliminated accordingly. Our future begins to sound like the opening of Woody Allen's parodic commencement speech, in which he writes that mankind faces a crossroads: one path leads to despair and utter hopelessness, and the other leads to total extinction. Let us pray we have the wisdom to choose correctly.

Actually the choices before us are not two but four: cultural studies and eventual doom-by-duplication; literary studies and eventual doom-by-irrelevance; cultural studies and prosperity with the nascent professional-managerial class; literary studies and prosperity with everyone who wants to study the aesthetic uses of language. And the way to decide which of these futures is most likely for literary study is simple: just figure out the exact relations among literary-slash-cultural study, cultural capital, student enrollments, and productive capital, and devise the curriculum accordingly.

I want to close this chapter by making a modest proposal, which, like all modest proposals, is much too radical for realization. It may be possible for English departments to foreground "literature" in the restricted belles-lettres sense in which the term has been understood in this century *at the same time* that literary study incorporates some of the concerns of cultural studies. It may be possible for English departments to make some claim on the interests of the professional-managerial class *even as* the needs of that class move further and further away from the Beowulf–to–Virginia Woolf canon of English literature. It may even be possible for English departments to expand their concern with the English language while becoming less . . . well, *English* in the process.

Notice if you will the odd fact that in debates over the place of cultural studies, critics rarely inquire into the place of contemporary writing in English. Even my own department's bylaws, as I noted above, consider "literatures in English" a thing apart from "cultural studies." All too often, when contemporary writing in English is taught (if it is taught at all) in literature departments, it comes under the heading of Postcolonial Literatures, World Lit, or (still worse) Third World Lit. In my own department, it is far and away the most underrepresented arena of writing; one of the stranger moments in our departmental deliberations, in fact, came just after some of my colleagues were finally convinced that "cultural studies" was a fair description of what some of us are doing at Illinois—when we *then* realized that we had advertised ourselves as offering "literatures in English" but only had two faculty members who had ever offered courses in literatures neither English nor American.

Reed Way Dasenbrock originally made this suggestion some years ago, before "cultural studies" had become a name for a fear and a desire, and I want to update his proposal and toss it onto the table just in time for the millennium. In Dasenbrock's terms, the literature curriculum is currently organized around a "centripetal" canon centered firmly in the British Isles—the canon bequeathed us largely by F. R. Leavis and *Scrutiny*, "from which," as Terry Eagleton has famously written, "criticism has never quite recovered" (28). This canon contains many of the greatest writers ever to inhabit and expand the English language, of course, but it also does double duty as an agent of Anglo-American national affiliation: just as the New Right likes to pretend that the United States has some deep genetic connection with Periclean Athens, so too do the Anglophile supporters of the centripetal canon like to pretend that you cannot understand "literatures in English" unless you have first completed the "coverage" requirements that will acquaint you with Gawain, the Miller's Tale, and MacFlecknoe—not to mention the three nineteenth-century British novelists who will secure your employment by Milton Rosenberg at WGN in Chicago. Yet it is not entirely clear, at the very least, that the British canon before 1790 (exclusive of Shakespeare, of course, who not only is our language's greatest writer but also cannot be challenged in the slightest without provoking a national scandal) is quite as deserving of the curricular place it now occupies in the United States; and it is not entirely clear why, if we now spend so much disciplinary time on British literature, we could not just as well (in some future incarnation) devote more of our time and energies to African, Indian, Caribbean, Australian, and Canadian writing in English.

One signal advantage of Dasenbrock's formulation is that it reshuffles but does not entirely jettison the centripetal canon with which we now work:

> In place of the centripetal canon oriented toward the England of Pope, Fielding, Richardson, and Johnson, a centrifugal canon might focus on Swift, Defoe, Smollett, and Boswell. This is a slight change, perhaps, as all these figures are recognized to be important and are placed in the canon somewhere. But as one moves closer and closer to the present, a centripetal, England-centered canon captures fewer and fewer of the important figures, whereas a centrifugal canon focused on the totality of writing in English has no difficulty at all in representing the

panorama of world writing in English. . . . Only a centrifugal conception of literature in English that embraces [George] Lamming and [J. M.] Coetzee as well as Shakespeare and Defoe can place all these works in the juxtapositions they deserve. (73–74)

Yet even as I cite these words I can hear the cries of my Augustan, Restoration, Renaissance/Early Modern, and medievalist colleagues, all of whom will insist that my modest proposal will marginalize their enterprises still further—at a time when most students are filing into their classes only because of their department's requirement that English majors take *x* number of credits in courses prior to 1800 exclusive of Shakespeare. Their concerns are entirely justified. Interestingly, some years ago when I taught Dasenbrock's essay as part of a graduate seminar on "canonicity and institutional criticism," my students almost uniformly dismissed Dasenbrock as too timidly reformist: his centrifugal canon would shuffle the deck a little bit, they allowed, but wouldn't alter the curriculum significantly enough to be called a strong form of canon revision. Politely but firmly, I pointed out to my students that our department contained almost no faculty qualified to teach contemporary world writers in English, and that Dasenbrock's proposal would therefore have the truly radical effect of compelling Illinois (and not Illinois alone) to alter completely its hiring policies in the future. It's one (relatively easy) thing to shuffle the canon, I suggested to my students; it's quite another (much harder) thing to shuffle the professoriate. The objection that the centrifugal canon would make the English department more "presentist," then, seems to me a valid point of concern—not least for those faculty who specialize in early periods of British literature.

To those faculty I can respond only by appealing to history, by way of John Guillory, who notes correctly that canon revision is—and has been for some time—a form of disciplinary "modernization":

The objective of canonical revision entails in practice shifting the weight of the syllabus from older works to *modern* works, since what is in question for us are *new* social identities and new writers. In fact, the history of the literary curriculum has always been characterized by a tendency to modernize the syllabus at the expense of older works. (15)

And if the history of the literary curriculum is in part a history of the relation between literary study and the distribution of cultural capital, well, then, if Buchi Emecheta, Salman Rushdie, Nadine Gordimer, and Michael Ondaatje will soon be displacing Pope and Dryden the way Faulkner and Fitzgerald did fifty years ago, perhaps our discipline will be all the healthier for it.

There is another plausible objection to this modest proposal; this one comes not from faculty in British literature before 1800 but from faculty who now teach world literature. A "literatures in English" paradigm, some might claim, represents one more attempt on the part of the so-called First World to shore up its economy by drawing on the resources of the so-called Third. In the case of literatures in English, the situation might arise that U.S. universities would be training American citizens to teach African literature, thereby maintaining one branch of the professional-managerial class (the professoriate) by providing them with raw material imported from "developing" nations. Certainly, a curriculum with expanded offerings in Canadian, Irish, Caribbean, African, or Australian literature would necessarily raise the question of whether U.S. colleges would hire a given number of Canadian, Irish, Caribbean, African, or Australian scholars as teachers. At the very least, such a curriculum would provoke the question of how and why literatures can travel across national boundaries more easily than can some literary critics. But this potential problem for the curriculum can also be a potential source of strength: whereas the original institutionalization of the literature curriculum, in the United Kingdom and the United States, had clearly served the purposes of nationalism and particularism, a "literatures in English" curriculum explicitly presents nationalism not as an assumption but as an object of interrogation—whether Joan Didion and Ngugi wa Thiong'o are meditating on the fate of national identity or James Joyce and George Lamming are exploring the politics of exile. It is all but impossible, I suggest, to make "literatures in English" into an instrument of national solidarity and self-definition, and the cosmopolitanism of such a curriculum, like its presentism, may be more of a benefit than it now appears.[5]

Yet the relation between "world literature in English" and "transnational cultural studies" is not straightforward, and I do not mean to introduce Emecheta et al. to the English curriculum simply in order to smuggle in some version of postcolonial cultural studies on the sly. No doubt it is possible to construct a curriculum of world literature in English around the airline-magazine proposition that because the planet is getting smaller every year, the enterprising professional-manager of the future will have to be acquainted with diverse global cultures if s/he is to open crucial new overseas markets. (Our intro course in the works of Amos Tutuola and Bessie Head will acquaint you with local African customs that will be invaluable for negotiating cultural difference to your advantage!) Nevertheless, it is hard to deny that a curriculum centered around world literature in English would necessarily bump up against some of the concerns of cultural studies, and it is entirely possible that cultural studies could be *realized* (in an institutional sense) all the more readily by an English department that is determinedly transnational in scope, attuned to and critical of the geopolitical imbalances of power that have made "world literatures in English" a subject available for literary study in the first place. If literary study wants to become more cultural, and cultural studies needs to become more literary, it is hard to imagine a more fitting institutional negotiation of these desires than an English department whose curriculum centers not on the British Isles but on the global ramifications of the world travels of the language first spoken on the British Isles.

There are many reasons I find this revised literary-slash-cultural studies curriculum to be compelling, but one stands out above all—the pragmatic one. At present, the profession does not have much of a public rationale for itself; that lack of a rationale, in turn, is both the condition and the product of the debate over the status of cultural studies. What rationale we have usually relies on our functions as teachers of writing— which is one reason the discipline of literary study can be said to be parasitic on the discipline of writing instruction, as composition theorists and writing studies faculty have been pointing out for some time. The writing-instruction rationale works not only because it is plausible (English departments do try to guarantee a certain kind of access to advanced literacy) but also, crucially, because it is politically *reversible:* it can be used to justify English as a discipline that fosters critical thinking at the same time it can be used to justify English as a discipline that fosters employability, business competence, and maybe even long-term financial security. As I'll argue in greater detail in chapter 6, the discipline has always required a rationale that has precisely this kind of political "reversibility"; and as much as I might hate to admit it as a progressive educator, and however much it might pain the liberals and conservatives to my right and the Marxists to my left, the rhetoric of public justification for intellectual work is necessarily a rhetoric of negotiation and double-voicedness—which is not at all the same thing, I hope, as a rhetoric of accommodation and double-talk.

But in the coming years, the writing-instruction rationale may very well leave English high and dry without a base of support for either literary studies *or* cultural studies. The reason is RCM, which in this context stands not for Royal Canadian Mounties but for Responsibility-Centered Management. RCM, apparently the latest successor to Total Quality Management, Theory Z, PQP, and myriad other business school budget-management techniques of the 1980s and 1990s, happens to be sweeping through the Big Ten universities of late, and the University of Illinois is in the process of converting to it as I write. In one way, RCM is an improvement over the opaque bureaucratic model now in place, for it has allowed us to discover that the budget of the College of Liberal Arts and Sciences contains far less than its share of state money, and that, conversely, agricultural enterprises at Illinois are major beneficiaries of the money LAS is not getting. But in another way, RCM puts the humanities permanently on the defensive: every college, every department, is allowed a kind of block grant the amount of which is fixed by formula, and the college or department can spend that money as it sees fit—for a new faculty member or for a new photocopier or for extended building hours (since everything pertaining to instruction, including the costs of the physical plant, is now the responsibility of the department or college). A department can generate revenue in one of four ways: federal/state support, franchise fees, private grants, or tuition. Since the humanities are not often on the receiving end of major public or private grant funds, and since the humanities, unlike the football team, sell no T-shirts or logo-ridden outerwear, departments like English or French will have to depend almost entirely on enrollment numbers for their financial support. And in English, those enrollment numbers—which will in turn generate the funds necessary for faculty and graduate students—will consist disproportionately of the warm freshman bodies processed by the university's required writing course.

In other words, in the era of RCM *every* English course, from Shakespeare's Great Tragedies and Romances to Que(e)rying the Beat: Gender, Sexuality, and Transgression in Postwar "Hip" Culture, will depend for its existence on the substrate of Intro Writing that keeps the department solvent. Insofar as the existence of English will be enrollment-driven, the debate over literary and cultural studies will inevitably turn into a debate over whether English is "surviving" or "caving in" by offering courses in contemporary culture. But it will be ultimately irrelevant to the *real* conditions of English, in which both traditional and untraditional forms of textual study will be just two kinds of dessert, more like than unlike, made available only so long as Intro Writing is offering a sufficient number of undergraduates the meat and potatoes of English—the only courses with enough status as cultural capital to induce large numbers of students and taxpayers to believe that they might just be worth the investment of some time and money.

It is in this context, finally, that we need to find a rhetoric of justification for literary study that incorporates some of the concerns of cultural studies *and* refrains from relying on introductory composition for public support. I do not believe that "contemporary world writing in English" answers all these concerns; surely, if John Guillory is right, the category of "literature" will not regain its prestige and cultural authority simply by adding exotic new writers to its lineup. All the same, I am surprised that so few of my colleagues—aside, obviously, from Reed Way Dasenbrock—have considered "literatures in English" as a possible organizational rubric, particularly since the literary-slash-cultural studies debate so often makes it sound as if we must rob Peter to pay Paul: the more culture, the less literature, and vice versa. "World writing in English," I think, at least holds out the appropriate prospect of making literary study more cultural, and cultural studies more literary, regardless of how many warm bodies are processed by Composition next semester.

To entertain this conclusion is to understand that in the university setting, disciplinary disputes, even of the most rarified kinds, are inevitably also disputes about relations of intellectual production. What we teach, and where we teach, affects how we hire; how we hire (intellectually as well as economically, from endowed chairs in cultural studies to adjuncts in introductory courses) profoundly affects what we teach. In the chapters that follow, then, I want to keep a dual focus on the employment of English—looking not only at how English can be employed (which will take up the bulk of the second half of this book) but also at the conditions of employment *in* English. And if we want to get a sense of the contemporary crisis in *those* conditions, we could not do better, at the present time, than to turn away from the competing claims of literature and culture for a moment—with the proviso that we will return to them shortly, but only after we have visited Yale.

NOTES

1. As Peter Brooks has said in his important essay, "Aesthetics and Ideology—What Happened to Poetics?"

   Students are only too willing to short-circuit the aesthetic, and to perform any kind of reading, including the ideological, that you indicate to them. What is more difficult for them—and hence more necessary—is to slow up the work of interpretation, the attempt to turn the text into *some other* discourse or system, and to consider it as a manifestation of the conventions, constraints and possibilities of literature. . . . Students need in the work on literature to encounter a moment of poetics—a moment in which they are forced to ask not only *what* the text means, but as well *how* it means, what its grounds as a meaning-making sign system are, and how we as readers, through the competence we have gained by reading other texts, activate and deploy systems that allow us to detect or create meaning. (In Levine 160–61)

2. One of the reasons it is useful to contest the narrative of decline is that the Right has jerry-rigged and publicized that narrative for its own purposes, and in so doing has distorted or suppressed everything that doesn't fit the narrative. For instance, the well-publicized National Association of Scholars' study of "core courses," released in 1996,

claimed that American universities had precipitously "declined" since 1964 in that fewer colleges required Western Civilization courses of their undergraduates. What the NAS study deliberately hides, however, is that although the number of universities offering core courses has declined since 1964, that number has *increased* in the past decade and a half; the reason the NAS chose 1964 as the basis of comparison, of course, is that it allows the NAS to blame the "decline" of higher education on their all-purpose scapegoat, the sixties. Pat Robertson and the Christian Coalition, in weaving their narrative of decline about primary public schools, like to focus on 1962, the year the Supreme Court banned prayer from the classroom; the NAS chooses 1964 presumably because once the Free Speech Movement had taken hold at Berkeley, the chant of "hey hey, ho ho, Western culture's got to go" was merely inevitable.

3. For discussion of the rise in graduate enrollments, by contrast, see Bérubé and Nelson 18–20.

4. For a more detailed version of my analysis of Readings, Guillory, and cultural capital, see my review essay, "The Abuses of the University," forthcoming in *American Literary History.*

5. For a discussion of "critical cosmopolitanism," see Robbins, *Secular Vocations;* for the relation between nationalism and canons, see Connor, *Postmodernist Culture;* Eagleton, *Literary Theory;* Graff, *Professing Literature;* and Shumway, *Creating American Civilization;* for "nationalist" literatures and their relation to postcolonialism, see Ahmad, *In Theory.*

2

THE BLESSED OF THE EARTH

In the fall of 1995, not long after graduate students at the University of Kansas voted to unionize, affiliating themselves with the American Federation of Teachers, I was invited to speak at Kansas on the future of graduate study in the humanities. In the course of my talk, I not only endorsed the unionization of graduate students at KU and elsewhere, but also referred, in passing, to what I called the "bad faith" attempt of administrators and faculty at Yale University to claim that their graduate students were simply students and not also "employees." As long as people are working as instructors or as teaching assistants and being paid for their work, I thought, it makes sense to consider them "employed," to consider their work "employment," and to admit, therefore, that they are in some sense "employees." And if administrators and faculty at Yale or elsewhere want to claim that their graduate students' wages are not "wages" because their teaching (which is not strictly "teaching") is merely part of their professional training as apprentice professors, then it makes sense to call the bluff: take graduate students out of the classrooms in which they work as graders, assistants, and instructors; maintain their stipend support at its current levels; and give them professional development and training that does not involve the direct supervision of undergraduates. *Then* we'll see how long Yale University can survive without the labor (which is not strictly "labor") of its graduate student teaching assistants.

At the time, I thought my support for graduate student unions—in a speech delivered to, among other people, unionized graduate students—amounted to endorsing candidates after they'd won their elections. To my surprise, however, I learned later that the graduate students were very pleased with my speech, and that some even considered it "courageous." It seems that I had denounced as ridiculous Yale administrators' claims that graduate students were not employees in front of a number of Kansas administrators who had claimed that graduate students were not employees. (I told the students I had had no idea that my audience included actual bad faith negotiators, and that my "courage" in denouncing them was therefore attributable to simple ignorance.) I asked them what other kinds of opposition the union had met; they told me of faculty in department after department who had insisted that the unionization of graduate students would disrupt "morale" and destroy the delicate, collegial relationship so characteristic of, and necessary to, healthy interactions between graduate students and faculty. When I asked these students whether their faculty had entertained the possibility that delicate, collegial relationships don't normally involve one party dictating the other party's interests and threatening punishment if party number two failed to act in what party number one had determined those interests to be, I was met with bitter laughter. It would be one thing, I was told, if the faculty's relation to graduate students were simply paternal rather than collegial; that would be undesirable but understandable. "But Michael," said one union leader, "half the faculty who spoke to us about the importance of faculty-student collegiality didn't even know our *names*."

Nothing, I suggest, could make more palpable the vast differences between Yale and Kansas. If there's one good thing we can say of the faculty who broke the graduate student strike at Yale University, it is this: they knew their students' names. Indeed, had they *not* known their students' names, they would not have been able to preserve the delicate, collegial faculty-student relationship at Yale by submitting their students' names to Yale's administration for disciplinary hearings and possible expulsion. As Yale president Richard Levin put it in a November 1994 letter to the chair of Yale's Graduate Employees and Students Organization (GESO), unionization of graduate students would inevitably "chill, rigidify, and diminish" the relationship between graduate students and their mentors and advisors on the faculty (qtd. in Young 180). Accordingly, from that point on, Yale graduate students who were not satisfied with their warm, flexible, and capacious relations with faculty members would have to be punished harshly and swiftly.

I will not attempt here to retell the history of graduate student organization at Yale, or the Yale Corporation's long and sorry history of union busting and unfair labor practices (for information on those histories, see Young, "On Strike at Yale," or contact Gordon Lafer, research director of the Federation of University Employees, the union with which GESO had voted to affiliate).[1] Instead, I want to examine a more narrowly professionalist issue—the role played by Yale faculty during the events leading up to the short-lived grade strike of 1995–96—and its implications for professional self-governance in American higher education. I believe the actions of the faculty at Yale have potentially grave consequences for the future of graduate study in the humanities and social sciences, just as they provide (less importantly but more poignantly) an object lesson in just how politically obtuse, shortsighted, and self-serving a university faculty can be.[2]

This is not to say that GESO has been always and everywhere beyond criticism, or that it is impossible for a well-informed person to lodge reasonable objections to the grade strike that precipitated the faculty's collective decision to crush GESO. That grade strike did indeed pit GESO against the interests of undergraduates and faculty alike, thus isolating the union politically and earning GESO harsh criticism from the Yale *Daily News,* the student newspaper. Moreover, it seemed at first to strain the meaning of "academic freedom" GESO had hoped would protect graduate students involved in job actions: to wit, GESO claimed that its members should be free from "academic reprisals, including letters of recommendation, disciplinary letters, academic probation, firing of teachers, denial of promised teaching jobs, or expulsion" (in the language of the resolution submitted by GESO to the MLA) and that any such action taken by Yale administration or faculty constituted a violation of academic freedom; but faculty responded that *their* academic freedom would be violated if they could not consider their students' participation in the grade strike as a factor in writing letters of recommendation or awarding teaching positions. It was not until November 1996 that a ruling by the National Labor Relations Board finally demolished the faculty position on their students' union activities: unambiguously, the NLRB held that Yale faculty who in any way penalized students for their involvement in GESO were in violation of federal law.[3] In the meantime, during the winter of 1995–96 when the Yale strike became national news, it was officially an unsettled question as to whether GESO's job actions were matters of labor relations or of academic protocol: if they were the former, then Yale was clearly involved in illegal union busting; if the latter, then striking GESO members were clearly abrogating one of their primary obligations as undergraduate instructors by failing to turn in their grades.

Of course, the grade strike made a crucial political point, a point that Yale's administration denied and incredibly *continues* to deny, namely, that a great deal of basic undergraduate instruction at Yale is carried out by graduate students. What's more, Yale students have convincingly argued that the strike was a measure of last resort; every prior attempt to meet and negotiate with Yale had been rebuffed. As Cynthia Young reports, by November 1995, the grade strike was the only effective action—short of a teaching strike—left to GESO. Demonstrations, petitions, a one-week strike, a union election, and corporation visits had all failed to convince the Yale administration that graduate teachers were indeed serious about winning a collective bargaining agreement. It was this bleak recognition that mobilized GESO organizers with barely three weeks left in the semester to begin organizing graduate teachers to withhold their grades. A grade strike would not only reinforce the central import of graduate teachers' labor at the university, but it would also undercut the Yale administration's attempts to depict GESO as dependent upon the other two locals to secure a contract. A grade strike barely a week before final exams had the capacity to spur undergraduates and faculty to pressure the administration to negotiate with GESO. It was certainly not intended as a strategy to harm undergraduates; in fact, striking teachers expressed their willingness to write letters to graduate and professional schools evaluating the student and explaining the reasons for the grade strike. In any case, it is unlikely that any school would have disqualified Yale candidates because of their incomplete transcripts. A grade strike is far less disruptive of undergraduate education than an indefinite teaching strike, a possibility that seemed to loom on the spring horizon. Weighing these various considerations, graduate teachers voted to withhold their fall semester grades until Yale committed to negotiating a written and binding agreement with GESO's negotiating committee. (188)

Even when considered in the light of these various justifications, however, the grade strike seems to have made two tactical errors in a Machiavellian sense. First, it underestimated the possibility that such an action would in fact spur undergraduates and faculty to pressure the administration to move forcefully against GESO. Second, and no less crucial, it regrettably allowed Yale faculty to pretend, after the fact, that they had been sympathetic to GESO, or generally supportive of graduate student grievances, or even opposed to GESO but in favor of collective student organization—until that deplorable grade strike came along and ruined everything.

The level of faculty vindictiveness and double-talk on this issue has been simply astounding. At various times, Yale faculty and administrators have claimed that they are opposed only to GESO and not to the idea of graduate student unionization; or that they are opposed to student unions at Yale but not other forms of collective (and nonbinding) student representation; or that they are opposed to unionization at Yale but not elsewhere, at other schools. It should not escape notice that each one of these rhetorical escape-maneuvers begs the original question concerning the sanctity of faculty-student relations. Perhaps it is plausible, for instance, that GESO would disrupt the delicate, collegial relations between graduate students and faculty, but another union would not. Or perhaps it is plausible that faculty would look kindly on graduate student representation that took some shape other than that of a union, as Peter Brooks has claimed.[4] Or finally, perhaps it is plausible that unionization always disrupts the faculty-student relationship, but does so in ways that can be tolerated at plebeian, inferior schools like the Universities of Kansas, Oregon, Michigan, Wisconsin-Madison, Wisconsin-Milwaukee, Massachusetts-Amherst, Massachusetts-Lowell, Florida, and South Florida, or at Rutgers, SUNY, and Berkeley (all of them home to recognized graduate student unions), but not at an institution so prestigious as Yale University, where talk of "unionization" is not only harmful to morale but also, and more vexingly, *bad form.*

What's remarkable is not that different Yale faculty have appealed to these various, contradictory rationales for union busting; what's remarkable is that *individual* faculty members have frantically appealed to each of them in turn, desperately trying to justify not only their opposition to the grade strike but also their intransigence during all GESO's attempts to negotiate prior to the strike. For a vivid illustration of this brand of double-talk I need only turn to my mailbox. On January 24, 1996, Annabel Patterson, Karl Young Professor of English at Yale, wrote a letter to Phyllis Franklin, the MLA's executive director, protesting the MLA Delegate Assembly's passage of the resolution censuring Yale for its handling of GESO. Patterson's letter, together with three other letters from Yale faculty and administrators, was circulated to the entire MLA membership in February 1996. There is much to remark upon both in Patterson's letter and in the manner of its distribution, but for now I want simply to focus on one crucial paragraph—the paragraph in which

Patterson addresses what she calls "the nature of the 'union'" (nowhere in Patterson's letter does she employ the terms "union" or "strike" without scare quotes). The reason the paragraph is valuable, for my purposes, is that it voices almost every single rhetorical escape-maneuver I enumerated above; when read together with Margaret Homans's equally evasive letter, also distributed by the MLA, it provides us with a useful introduction to faculty psychology at Yale.

Patterson writes,

> The university administration, whose leaders are all Yale faculty, has consistently refused to recognize [GESO] as a union, not only because it does not believe this to be an appropriate relationship between students and faculty in a non-profit organization, but also because GESO has always been a wing of Locals 34 and 35 of the Hotel Employees and Restaurant Employees International Union, who draw their membership from the dining workers in the colleges and other support staff. Yale is not prepared to negotiate academic policy, such as the structure of the teaching program or class size, with the Hotel Employees and Restaurant Employees International Union. Yale administrators have made it perfectly clear that they have no objections to working with an elected graduate student organization other than GESO, one that is not tied to the non-academic unions on campus. (6)

According to Patterson, Yale has properly refused to recognize GESO because the graduate student "union" is affiliated with the smelly hotel and restaurant workers, who don't know how a university works. But wait a minute: look at the closing and opening passages of Patterson's paragraph. Apparently, Yale has no aversion to "working with an elected graduate student organization other than GESO" so long as the organization is not tied to Locals 34 and 35. Does this mean that Yale would have been happy to recognize GESO if only GESO had had the good taste to affiliate with the AFT? The earlier passage had seemed to close off this possibility, declaring that Yale has refused to recognize GESO as a union because "it does not believe this to be an appropriate relationship between students and faculty in a non-profit organization." So what is one to conclude from this? If only GESO hadn't affiliated with a nonacademic union . . . if only GESO had been something other than a union . . . and (by the bye) if only the Yale Corporation were something other than a nonprofit institution . . . *then*, obviously, Patterson implies, we'd have had no objection at all to dealing with these students in good faith.

Margaret Homans then adds two more "if" clauses to this already impressively obfuscatory list when she writes, in her January 14 letter to the MLA,

> Quite possibly, it would be appropriate for students to unionize at those schools where teaching loads are much higher than at Yale and where reliance on graduate teaching is much greater. Part-time and adjunct faculty with Ph.D.s present an even more legitimate motive to unionize, although they are not part of the union movement at Yale. (11)

If only they were worse off, like those students at second-rate schools like Berkeley . . . if only they were among the *truly* exploited, like part-time and adjunct faculty . . . why, of *course* we would break bread with these students. Note here that Homans's admission that graduate student unionization is sometimes appropriate (at lesser schools) makes hash of the claim that faculty-student relations are destroyed by unions. Yet Homans's attempt to play one underpaid constituency off another—in this case, juxtaposing graduate students to adjuncts—presents an odd mixture of fuzzy thinking and bad faith: fuzzy thinking, because adjunct faculty already have the right to unionize (precisely the right denied to Yale's graduate students), and bad faith, because the nation's largest union of college faculty, the AAUP, had already disposed of this question, when its Collective Bargaining Congress passed a resolution on December 2, 1995, strongly endorsing the right of *all* graduate teaching assistants to engage in union activities, from collective bargaining to grade strikes.

It is possible that somewhere deep in the recesses of its political unconscious, Homans's text always already acknowledges its bad faith in adjudicating and ranking the rights claims of graduate students and adjunct faculty; for no sooner does Homans mention the exploitation of adjuncts than she moves on to threaten Yale students with the exploitation of adjuncts. "The students who introduced the resolution," she writes, referring to the MLA Delegate Assembly's resolution to censure,

> captured and capitalized on a legitimate anxiety, widespread in the profession, about the exploitation of non-ladder instructors. But graduate students at Yale are "paid" more (in some cases twice as much) for running a weekly discussion section of a lecture course (often with as few as fifteen students) than Ph.D.s are paid for teaching their own independent courses at area schools. . . . If they were paid the local rate for part-time academic work, they would receive a good deal less. (11)

What is the implication of this last sentence? Take *that,* you pampered, sheltered students! You people haven't yet *seen* what we could do to you if we really wanted to *exploit* you! If Patterson's letter was notable for the extent of its author's identification with the Yale administration—"Yale is not prepared to negotiate academic policy . . . with the Hotel Employees and Restaurant Employees International Union"—then Homans's is notable for its author's willingness to begin the union busting herself. For why else would Homans remind Yale graduate students (as if they needed to be reminded) that Ph.D.s are working for even lower wages at the University of Bridgeport or Southern Connecticut State? (Though Homans does not acknowledge as much, rumor has it that the endowments and budgets of Bridgeport and Southern are somewhat smaller than Yale's.) Is GESO, then, supposed to be grateful that their masters and overseers at Yale are at least treating them better than the freeway fliers at the college down the road? "Well," one imagines a Yale ABD replying, "we're paid $2,000 less than Yale's own cost-of-living estimate for New Haven, and Yale requires that we live here so that we cannot seek higher-paying part-time employment elsewhere while pursuing our degree; but golly, it's great that we're doing so well compared to the part-time schleps and losers at New Haven's own Albertus Magnus College, a nearly penniless institution. Thank goodness Professor Homans straightened us out on that one."

Despite the passages I've cited above, Homans's letter is not unaware that unethical labor practices might in fact be unethical. Though Homans is not shy about suggesting that graduate students be paid "the local rate" for discussion sections in which they do all the grading (so that people like Homans don't have to), she is appropriately uneasy about the charge that Yale might have had plans to hire "replacement workers" to take on the teaching responsibilities of striking graduate students when classes resumed in the spring of 1996. The aura of hiring "replacement workers" is apparently more unsavory than the aura of breaking unions and depressing wage scales, and thus Homans writes,

> The most basic standards of evidence were not adhered to in the formulation of the resolution, which complains (for example) of faculty being asked to "serve as replacement workers for striking graduate student staff." Faculty teaching lecture courses are in fact responsible for all grades; forms for reporting grades are mailed only to the faculty in charge and not to the Teaching Assistants, who are exactly that—assistants. We can't be described as replacement workers if we turn in grades for our own courses. (10)

One has to admire the faculty member who can write this without fear of exposure or contradiction. *Faculty are responsible for all grades:* the wording suggests that Yale faculty are actually reading the papers and evaluating the written and oral work of all their undergraduates, when, in fact, teaching assistants in lecture courses are hired precisely to release faculty from much of the labor associated with those tasks. (Hence the rationale for the grade strike.) One wonders how many MLA members, many of whom are actually college faculty themselves, could possibly be fooled by Homans's reasoning here: *the grade forms are mailed to us and not to the "assistants," so obviously we're the ones doing the grading!*

Delectable also is the "we" in Homans's declaration that "we can't be described as replacement workers if we turn in grades for our own courses." For one thing, the fear at Yale was not that Professors Homans and Patterson would step in and teach extra classes; the fear was that *junior faculty*—who you mean, "we"?—would be "asked" to teach in place of graduate students, or, still more outrageously, to do the grading for the lecture courses of senior faculty (some reports indicate that this latter request was in fact made by the senior faculty of the English department). And for another thing, Homans's letter is in this respect directly contradicted by Patterson, who admits freely that "some classes had been reassigned to faculty members" (5). (Personally, I am glad that Yale faculty have so little practice in conducting disinformation campaigns. Were they more practiced at the art they would never have let a major slip like this get into a mass mailing.) Homans, of course, would countercharge that faculty can never be considered "replacement workers." Again, though, one wonders who might be fooled by this. Even if faculty turn in all the grades "for their own courses" (once their teaching assistants have collected them, that is), that doesn't mean that faculty are not being used as replacement workers when they are asked to turn in the grades for *other people's* courses, particularly when those other people are out on strike. A faculty member who is asked to teach a course or lead a discussion section for a striking graduate student is being asked to cross a picket line, and thus to serve as a replacement worker. That should be clear enough. And when the faculty member in question is untenured, then such a request broaches serious ethical and professional issues that neither Homans nor Patterson attends to. That, too, should be clear enough.

Yet why is it *not* clear enough to most of the senior faculty most immediately involved? I want to suggest that something strange is going on here. When a professor of English begins sounding like an employer of migrant citrus workers *(at least you're being paid here—at Sunkist they give their workers only an orange a day),* or when the possessor of a named chair at one of the world's wealthiest universities insists that $9,750 is more than adequate compensation for graduate teaching assistants (see Patterson 6), then clearly some of the protocols of the profession have gone haywire. For the response of the Yale faculty to GESO is by no means confined to the rhetorical circumlocutions of Homans and Patterson; on the contrary, as Patterson herself notes, a special late-December meeting of Yale faculty, attended by 170 persons, indicated "*overwhelming* support for President Levin's policy of refusing to recognize GESO, with perhaps half a dozen voices against it" (7; emphasis in original)—and Michael Denning, one of those half dozen voices, does not dispute the numbers. David Brion Davis, professor of history, went a good deal further than Homans or Patterson, and submitted the name of one of his students, Diana Paton, to the office of the dean for disciplinary hearings, as did Sara Suleri-Goodyear, postcolonial critic extraordinaire (in the case of Cynthia Young);[5] meanwhile, Thomas Carew, chair of the Department of Psychology, called one of his students in India during the winter break, "falsely informing her that everyone else in the department had dropped out of the grade strike" (qtd. in Gage 11). Some faculty, it appears, were truly eager to go the extra mile to break the strike and punish the students they "mentor."

But the full extent of the group psychosis involved in these faculty responses to GESO doesn't begin to come clear, I think, until you step back and realize that for all their bellowing and blustering, *Yale faculty had no direct stake in the prospect of unionization.* GESO was not demanding to have their salaries augmented by stripping Annabel Patterson of the Karl Young chair; at no time did GESO demand that David Brion Davis be personally prevented from dictating university policy regarding class size and health care for graduate teaching assistants. Nevertheless, many Yale faculty insisted that graduate student unionization would take fundamental issues concerning graduate employment out of their hands, apparently oblivious to the fact that most of the issues GESO had placed on the table—from salaries to health care—were always already out of their hands. Faculty resistance to GESO, then, was almost entirely a matter of imaginary relations to real conditions, as Peter Brooks amply demonstrated when he claimed that "a union just seems to militate against core values" (qtd. in Eakin 58).

No commentator on the Yale strike has yet made this most obvious point: until the grade strike, Yale faculty had nothing important to lose in recognizing GESO. By contrast, once the grade strike was under way, *then* Yale faculty most certainly had something material at stake, namely, public recognition of the fact that graduate students do more hands-on teaching and evaluating of undergraduates than faculty do.[6] One would think that any sane, calculating university faculty members who are interested in maintaining their privileges and hierarchies—and few faculties, clearly, are so interested in this as are Yale's faculty—would have foreseen the potentially explosive political ramifications of well-publicized job actions by graduate students, and moved to palliate GESO with band-aid, stopgap measures while the faculty still had nothing at stake in the dispute. The fact that the faculty did not do so suggests that we should not look for "real" explanations of the Yale dispute—we should look instead to the realm of the Imaginary.

By their own report, antiunion faculty at Yale were stunned by the volume of GESO's sympathetic support among faculty members at other institutions; hence their obsessive insistence on their own near unanimity in opposing the grade strike, and their willingness to accuse GESO of lying in order to manipulate public opinion. As Annabel Patterson puts it, when Yale received over three hundred letters from faculty protesting Yale's refusal to recognize GESO, "we observed that many of [the letters] were from people conscious that they were hearing only one side of the story" (7). In other words, GESO's external supporters (including myself) were really rather tentative, because they knew they had not yet taken into account the weight (and the prestige) of the opinions of Yale's senior faculty. The level of arrogance here is audible. But if you want to get a vivid sense of just how insular and blinkered Yale's senior faculty have been with regard to the broader issues at stake in the recognition of GESO, Patterson's letter is insufficient on its own; you need to hear another side of the story. You need, at the very least, to read an account of the Yale strike written by people for whom the legitimation crisis of American higher education is always foremost on the agenda:

> There can be little doubt that graduate students at Yale, like graduate students almost everywhere, are exploited as cheap labor. Teaching assistantships are notoriously poorly paid, and the rationale that they should provide a welcome "apprenticeship" for future college professors looks more and more shabby as universities increasingly rely on these cadres of relatively untrained teachers to supplement their regular professorial ranks at discount prices. In fact, Yale has been better than most institutions at requiring its "big name" professors actually to teach undergraduates. But even at Yale, *the habit of fobbing off the ever*

more expensive education of undergraduates on teaching assistants is a scandal waiting to be exploded. For graduate students, teaching has more and more become simply a form of financial aid instead of a genuine apprenticeship; for universities, graduate students have become more and more like a pool of migrant workers. (3; my emphasis)

There isn't a false note in this passage, but you'll search in vain for this succinct, scathing analysis of American universities' labor relations in the pages of the *MLA Newsletter*. It appeared, instead, in that stalwart voice of trade unionist activism, the *New Criterion*.

Of course, the folks at the *New Criterion* have only a limited sympathy with GESO, and the unsigned editorial goes on to inveigh against the existence of *any* university-based unions, not only among graduate students but also among faculty, claiming incoherently that "the idea that students of any description should seek to organize themselves into a union is preposterous. The spectacle of graduate students doing so is only marginally less ludicrous than the prospect of undergraduates or high-school students doing so would be" (3). Somewhere between paragraphs, surely, the *New Criterion* editors forgot that graduate students teach classes whereas undergraduates and high school students generally do not; and you would think Roger Kimball, managing editor of the *New Criterion*, would have good reason not to forget this, since he himself taught undergraduates at Yale when he was a graduate student in the English department at the turn of the eighties (such was the basis of the claim on the back of his famous book, *Tenured Radicals*, that he had once taught at Yale). But whatever the source of the *New Criterion's* schizophrenia concerning graduate teaching assistants, one thing is indisputable: when the editors of the *New Criterion* have a vastly better sense of what's at stake at Yale than the faculty at Yale, it's time for some serious perestroika in the groves of academe. Yale officially insists, of course, that all its "teaching fellows" are guided and supervised by a faculty member, but this claim is emphatically contradicted even by one of GESO's strongest critics, Camille Ibbotson, who told *Lingua Franca* not only that "no faculty member has ever visited my class or expressed an interest in what I was doing" but also that "there is no formal teacher training in my department" (qtd. in Eakin 60).

Surely, part of this debacle is attributable specifically to pathologies endemic to Yale and Yale alone. The Yale corporation has long had a history of toxic aversion to unionization of any kind, be it among graduate students or clerical workers, and the vast majority of Yale faculty, apparently fully interpellated as members of the Corporation, seem to have such an enormous investment in their own prestige that the very idea of unionization threatens their sense of privilege, their sense of *distinction* from mere public universities like Kansas and Berkeley. The weight of "prestige" in the collective faculty imaginary should not be underestimated here. The *New Criterion* casts Yale graduate students as "exploited cheap labor"; Peter Brooks insists that "they really are among the blessed of the earth" (qtd. in Eakin 56). They are not, after all, just any garden-variety cheap labor; they are cheap labor *at Yale.* What makes Brooks's insistence all the more interesting is that Brooks is reportedly one of the few anti-union Yale faculty who freely admit that TA teaching loads (in contact hours) have risen over the past twenty years while wages (per hour, adjusted for inflation) have fallen. That profile sounds more like the plight of post-Fordist American workers in general—higher productivity, lower wages—than a description of the blessed of the earth. Does Brooks know a secret the *New Criterion* and the AAUP do not know? Or is Brooks revealing something about the assumptions undergirding graduate instruction at Yale?

Let me propose the latter, and let me further propose that if I am right, then many Yale faculty may have been not merely offended but positively *hurt,* emotionally and professionally, by the existence—and the persistence—of GESO. When Yale graduate students point to the job market as evidence that humanities Ph.D.s are not automatically to be classed among the blessed of the earth, what must this argument signify to Yale faculty? The very premise of the school is that there is no need to pay graduate students a "living wage," since the Yale degree assures them of lucrative academic employment at the end of their term as "apprentices." When Yale students reply to this premise by pointing

THE BLESSED OF THE EARTH

out their school's abysmal placement record in the humanities, what are they saying? They're saying that Yale is not exempt from the rest of the economy in American higher education. They're saying that they're not the blessed of the earth, any more than are the graduate teaching assistants at the University of Kansas. And *that* means that Yale faculty are no longer so uniformly powerful as to grant their Ph.D. students exemption from the great depression in the academic job market.

Recall that Yale has more to lose than most schools in this respect, particularly with regard to the self-regard of its faculty in the modern languages. It was not long ago that Yale was not merely a school but a School, where protégés and epigones could be produced in the high European manner, carrying forward the work of the Yale masters in learned journals and even (sometimes) in the interior of the continent. Back when Roger Kimball was still working away at his dissertation, Yale dominated the English charts in the manner of the early Beatles, and Paul, J. Hillis, Geoff, and Harold "Ringo" Bloom made their insights and influence felt even as they redefined "influence" and "insight." Later came the breakup, the solo efforts, the persistent rumors that Paul was dead. But all that did not matter, because the imprimatur of the Yale degree was still a sure thing, academe's version of a vintage Lennon/McCartney single. If GESO has done nothing else, the union has put Yale faculty on notice that this is no longer the case. And the revelation is so painful, it seems, that the vast majority of affected faculty can respond only by lashing out at the students who would dare to act on the recognition, *pace* Homans and Brooks, that graduate student labor at Yale is not, in the end, significantly different—even after the Ph.D. has been granted and the years of "apprenticeship" ostensibly ended—from graduate student labor at Kansas.

In one sense, then, Yale is an object lesson only for Yale. But in another, more important sense, Yale is not a special case at all; on the contrary, the events at Yale in 1995–96 might very well signal a new day in higher education throughout the United States. Toward the end of her letter to the MLA, Margaret Homans names the problem precisely, arguing for Yale's exemption from the academic economy in terms that make clear why Yale is not exempt from the academic economy: "I believe the delegates [who voted to censure Yale] confused legitimate problems in academic labor relations with issues quite specific to the situation at Yale, issues of which they seemed content to remain ignorant. . . . The exploitation of academic professionals—a national problem—is being trivialized for the sake of winning a small, elite group a fleeting PR victory" (11). In a dazzling display of looking-glass logic, Homans has derived exactly the wrong lesson from the job actions at Yale: her argument is not only (once again) that there are real problems *elsewhere* that have no bearing on the blessed graduate students of Yale; now, her argument is that GESO, by highlighting the "national problem" of exploited academic professionals, by putting the issue in the pages of major American newspapers up and down the Eastern seaboard, has somehow *trivialized* the problem. Thank goodness the *New Criterion* knows better: the exploitation of academic professionals is indeed a national problem, and Yale is but the leading edge of a national scandal.

Think of Yale this way: the university's endowment is already well over $4 billion, and recently has been growing faster than the national debt. According to a document released over the Internet by Michael Denning, "The University's investments manager recently revealed that Yale's endowment is having its best year in a decade. In 1995–96, the endowment will earn roughly $1 billion—after accounting for all expenses, Yale is earning almost $2 million a day, every day of the year." Moreover, whatever the limitations of its humanities faculty, the school remains relatively well respected and much in demand among high school graduates (though one presumes that aspiring graduate students in the modern languages, if they have some sense of self-preservation, will want to apply elsewhere in the future). Given Yale's extremely fortunate position in American academe, then, it should not have been hard for Yale faculty to have adopted something like the following reasoning: *if Yale University can't pay graduate students a living wage, complete with free health care, then who can?*

The reason so few Yale faculty have adopted this reasoning, I suggest, is precisely that they cannot see any structural relation between Yale and the vast legions of lesser American schools. The idea, for instance, that destroying GESO at Yale might just have deleterious effects for graduate student unions elsewhere (even at schools where such things might conceivably be necessary) seems never to have occurred to Homans or to her colleagues in arms. Likewise, none of GESO's opponents on the Yale faculty seems even to have entertained the possibility that other universities might look to Yale and say, "If a school so incredibly rich can farm out so much of its undergraduate instruction to adjuncts and graduate students, surely we have all the more reason to rely on part-time labor." Nothing, I submit, could be more painfully indicative of academe's *idiot savant* culture than the spectacle of dozens of bright, articulate scholars, skilled at reading mediations, overdeterminations, and cultural texts galore but incapable of understanding that their relations to graduate students at their own university might just have repercussions for labor relations at other universities.

As if this spectacle weren't depressing enough, there's the further question of GESO's relation to Locals 34 and 35 of the Hotel Employees and Restaurant Employees International Union. Here I must shed the temperate language I have used to this point, and speak bluntly for a change: in late 1995 any damn fool, even a distinguished Yale professor, could have seen that the Yale administration's attempt to crush GESO was but the prelude to its full-scale attempt to crush Locals 34 and 35 in the spring of 1996. Yale faculty may have been offended that their doctoral students had chosen to consort with menial laborers, but Yale administrators had a much better reason to oppose the affiliation: recognition of GESO would have complicated—perhaps even short-circuited—their plans to devastate the working conditions of Yale employees across the board.

Here, in a nutshell, is what those plans look like. One of the world's wealthiest universities proposes to cut future workers' wages by 40 percent and redefine them as ten-month workers so as not to pay them benefits. Again, this is at a school that's clearing a cool $2 million a day. As Denning's Internet communiqué noted, "Since Yale is realizing this level of profit under the current labor contracts, it cannot be that drastic cuts are required for the university's fiscal health." The Yale labor pool is (of course) overwhelmingly nonwhite and drawn from New Haven, the seventh poorest city in the United States; Yale is by far the city's biggest employer, accounting for roughly one in seven city jobs. According to Gordon Lafer of FUE, when Locals 34 and 35 went out on strike, during one of New Haven's coldest winters on record, the university tried to ban workers from keeping fires in oil cans for warmth on the grounds that the fumes would violate campus air quality standards; when a local bakery offered its day-old bread to striking workers, Yale threatened to cut off all future contracts with the bakery unless the bread was thrown out. Yale's new policies for its service staff are so draconian and mean-spirited, in fact, that I do not know whether to call them post-Fordist or pre-Fordist. So let's simply call them obscene.

Annabel Patterson's letter to the MLA, as I have noted, remarks that the leaders of the Yale administration "are all Yale faculty"; presumably Patterson made this point in order to suggest that she and her colleagues were professionally bound to stand by their men in their opposition to GESO. The question for Patterson and her colleagues, then, is this: does that logic also dictate that Yale faculty should support their administration's Dickensian assaults on the workers in Locals 34 and 35? Financially there is absolutely no justification for Yale's latest effort at union busting: the university is rich and getting richer, an enviable position for a nonprofit institution. One would think, therefore, that Yale's senior faculty, being the humane, decent people they are, would oppose their administration's policies with regard to Locals 34 and 35. But then, one would also have thought that Yale faculty, being the smart, well-spoken people they are, would have seen the connection between their university's opposition to GESO and their university's broader plans for union busting on campus.

If ever an institutional crisis demanded the attention of professional organizations like the MLA, this is it. But the MLA's response to the strike at Yale was somewhat less than encouraging. Six weeks after the Delegate Assembly passed the resolution censuring Yale in December 1995, the MLA conducted its mass mailing of the letters of Homans, Patterson et al., introducing its twelve-page document with the words "we write to initiate a new procedure" (1). The chief purpose of the mailing was to circulate to the MLA membership the views of Yale faculty opposed to GESO, the grade strike, and the resolution. No views sympathetic to GESO were included. In subsequent communications, the rationale for the mailing became clear: the GESO forces had had their say during the MLA convention, and, according to Margaret Homans, Yale faculty had not been able to respond sufficiently to the resolution at the time it was proposed: "if the MLA sees itself as representing and honoring diversity of opinion," Homans wrote, "the process by which the resolution was pushed through gives the lie to that claim" (10). (Homans and Brooks were both present at the Delegate Assembly, though Homans's letter does not indicate as much.) The MLA staff dutifully and rigorously investigated the charges that the resolution had been improperly introduced, and found, in the words of executive director Phyllis Franklin, that "the assembly's action was valid" (1). So much for Homans's precarious sense of proper procedure. Nevertheless, the mailing itself quite clearly seems to accept Homans's charge that "diversity of opinion" was not honored at the convention; no other explanation will account for the MLA's curious decision not to seek opinions sympathetic to GESO for the purposes of the mailing. As a result, the claims of Yale faculty were allowed to stand utterly uncontested—including Homans's unsubstantiated and grossly misleading "procedural" complaints that "the most basic standards of evidence were not adhered to in the formulation of the resolution" (regarding the status of faculty as "replacement workers") and that "the resolution violates several of legal counsel's criteria for acceptable resolutions: it is factually erroneous, slanderous, and personally motivated" (10).

When I first read over the special MLA mailing, I was appalled—so appalled that I did not consider it worth my time to complain to the MLA directly. Instead, I considered leaving the organization altogether. A great deal of effort and deliberation had obviously gone into the production and mailing of this unprecedented and one-sided document; a portion of my MLA dues had supported it, as had a portion of the dues of every graduate student and adjunct faculty member in the MLA; and as a result, my own professional organization had clearly given its members the strong impression that the Yale resolution was ethically dubious and factually mistaken. Ironically, Homans's claim that the

MLA had violated its commitment to "diversity of opinion" had been circulated to over thirty thousand faculty and graduate students without a single word of rebuttal; the claims of Yale faculty that the Yale resolution was ethically dubious were themselves circulated in an ethically dubious manner. It is testimony to the outrageousness of faculty behavior at Yale—and testimony to the air of unreality with which Yale faculty spoke of the self-evident rightness of their behavior—that *despite* the MLA mass mailing, the MLA membership voted convincingly, 3,828 to 2,474 (with 836 abstentions), to uphold the Delegate Assembly motion to censure.

Yet at the very least, the MLA mailing suggested that when confronted with a professional dispute between senior faculty and graduate students, the organization would go to extraordinary lengths, even "initiate a new procedure," to publicize the views of senior faculty *at the expense of* the views of graduate students. It is worth remembering here that the Yale resolution is one of the few substantive resolutions the MLA has passed in many years that materially addresses the professional working conditions of MLA members; the other burning issues on the table for 1996, for instance, included a resolution expressing "appreciation of and respect for the support staffs in our departments" and another resolution recommending a "common application form" for fellowships in the humanities. It is difficult, in the wake of the MLA's February 9 mass mailing, to imagine what the professional role of the MLA—and its Delegate Assembly, to which I was elected in the spring of 1996—can conceivably be. For the moment, it appears that the MLA is quite efficient at passing resolutions about being nice to secretaries, treating books with extra care, condemning U.S. foreign policy, and refusing to hold the national convention in forty-six of the fifty states. But when the MLA at last confronts an issue that addresses head-on the crisis of labor relations in American universities, the entire "resolution" system is thrown into profound crisis—by, of all things, the objections of a small handful of elite faculty seeking to win a fleeting PR victory.

And yet if recent *MLA Newsletters* are any indication of the state of the profession at its highest echelons, MLA inattention to academic labor relations may prove to be much less harmful to the profession than actual

MLA *attention* to academic labor relations. In the winter of 1995, as the Yale standoff heated up and thousands of new and recent Ph.D.s made their preparations to attend the MLA convention for yet another costly and generally fruitless exercise in job hunting, the *MLA Newsletter* featured a column by the brilliant and internationally renowned Sander Gilman, who, writing his final column as MLA president for 1995, proposed a novel solution to the job crisis in the humanities. The column, "Jobs: What We (Not They) Can Do," was written explicitly as a response to angry graduate students caught in the job crunch. Gilman opens by narrating a confrontation with such graduate students at the 1994 MLA convention, remarking that "it was clear that the candidates' anger was directed not at any amorphous 'they' but at their own professional organization, the MLA, and that they were yelling at me not because I had done anything specifically to block them from getting jobs but because I represented that force of nature, the MLA—that is, 'us' " (4). He proceeds thence to suggest that the MLA create "postdoctoral mentored teaching fellowships—nontenured, two-year appointments with limited benefit packages" (4). These mentored postdocs, writes Gilman, will solve the profession's employment crisis by offering younger colleagues "serious, meaningful employment" (5) while also affording "the flexibility administrators demand in our fields" (4).

One can only guess at what "flexibility" might mean here (it seems to be a synonym for "fire-ability"), let alone why "flexibility" might be an employment criterion that a professional organization like the MLA would seek to embrace. Gilman notes, in a brief remark uncannily like that of Annabel Patterson's insistence that Yale's leading administrators are also Yale faculty, that his plan will be smiled upon by those above: "we can create new jobs in our departments if our administrators, many of whom are also members of the MLA, see that we are serious in our desire to reallocate resources" (4). In other words, our administrations are downsizing, but *they* are really "us"; the graduate students who were once part of that "us," in an earlier paragraph, are now resources to be reallocated so that "we" can show "our" administrators how serious we are about signing on to the latest downsizing initiative. Gilman briefly suggests that his proposal is a kinder, gentler form of exploitation—

"new Ph.D.s will become better teachers," he suggests, as if they haven't already done enough teaching as graduate students, and "faculty members will have rewarding mentoring tasks" (4). But what if the senior faculty don't want to "mentor" these two-year, part-time, piecework pseudo-colleagues? No problem, says Gilman—we'll just leave out the "kinder, gentler" part: "if we [note the "we" here] don't want to take on a mentoring role because of our overloaded schedules, we can create two-year lecturer positions" (5).

What follows this bizarre suggestion is a still more bizarre paragraph insisting that we should not hire undergraduates as unpaid laborers to teach "drill sections." "Nor should we listen," continues Gilman, "to the argument that this arrangement provides a perfect apprenticeship for students who plan to go to graduate school" (5). Here, I think, is an "argument" beyond human comprehension: who, exactly, is arguing that we should staff undergraduate courses with undergraduate teachers as "apprentice" graduate students? I cannot answer this question, but I can suggest that Gilman's stern, forceful paragraph ruling out the use of undergraduate instructors serves the purpose of making his own mentored-postdoc suggestion sound "reasonable" by juxtaposing it to the truly insane option of having undergraduate classes taught by unpaid undergraduates. There's nothing wrong with creating a new tier of second-class faculty, in other words, but when it comes to charging undergraduates tuition to teach themselves in drill sections, *that* we will not countenance.

For what, in the end, is Gilman really proposing, and how would it work? In his antepenultimate paragraph, he writes, "Graduate programs that still admit masses of graduate students could temporarily amalgamate two teaching assistantships into a two-year postdoc. Institutions would receive the same amount of teaching for less money, because they would not have to pay graduate school tuition for these postdoctoral fellows" (5). Let's parse out this suggestion carefully. Apparently, Gilman's postdocs would teach at twice the pay scale of graduate teaching assistants, and teach twice the course load, thus providing their institutions with the labor of two graduate students. All right. At Illinois, that would mean that the Sander Gilman Flexible Postdoctoral Fellows would earn just over $21,000 a year for teaching four courses per semester. And, Gilman adds, Illinois would not have to pay their tuition. But of course, Illinois does not "pay" the tuition of any graduate student; it *waives* graduate student tuition in return for undercompensated teaching (and even that arrangement is being contested as I write). No money changes hands in a tuition waiver; the transaction happens entirely in an executive assistant's software program, as spreadsheet numbers are fiddled and adjusted. The idea that universities "pay" their graduate students' tuition, in other words, is an especially threadbare fiction, though it seems to have been put to good use by the anti-GESO faculty at Yale, who are apt to claim that their students are "paid" almost $20,000 yearly in tuition waivers—as if the university is gallantly taking a loss by providing graduate students with $20K worth of valuable instruction at no charge. It is this threadbare fiction that allows Gilman to present his plan as a money-saver ("the same amount of teaching for less money"), as if universities actually gave tuition waivers in cash, and could pocket the dollars themselves by hiring a Gilman Flexible Fellow.

Tuition waivers, however, are not the crucial issue for Gilman's argument. The crucial issue is that if Gilman's argument becomes widely circulated in American universities, the profession of college teaching as we know it is basically finished. "Let us generate new postdoctoral fellowships throughout the country," writes Gilman (5). Lethal as this might be to the future of tenure-track employment, in some ways it is not a bad idea: if the going rate for these Gilman Fellows is $2,500 per course, many of my former students, teaching at small colleges as part-time laborers, are in for a raise of anywhere from 60 to 200 percent. But I don't think that's going to happen. The colleges that now employ Ph.D.s at the rate of $800 to $1,500 per course are not likely to sign on to the Gilman Program in order to convince senior administrators of their "seriousness." (And, I should add, Ph.D.s who teach at these rates are extremely unlikely to need further "mentoring" to hone their pedagogical skills.) For many American colleges, then, Gilman's proposal is simply irrelevant. What then of the colleges that now employ Ph.D.s as assistant professors, at the rate of $30,000 to $40,000? Wouldn't they do well to cut their salary and benefit costs by eliminating tenure-track faculty entirely and hiring, instead, new Gilman Fellows with limited benefit packages? For such colleges, I cannot imagine a labor relations "solution" more administration-friendly than Gilman's. If you want a flexible workforce at a discount rate, there's no need to mount difficult, costly legal challenges to the institution of tenure; just hire a gaggle of part-time Gilman Fellows at $20,000 with optional health coverage (mentoring also optional), and presto, you've created a new stratum of part-time faculty while saving your institution untold thousands of dollars in salaries and benefits. And *that* will show you're serious in your desire to reallocate resources.

What Gilman is proposing for new Ph.D.s, in other words, is precisely what Yale is proposing for Locals 34 and 35: a 40 percent pay cut (from $35,000 to $21,000, more or less), redefinition as part-time labor, and a significant rollback in benefits. For some reason I do not understand, Gilman seems to believe that university administrators will agree to create a wholly separate category of underpaid, part-time, short-term faculty *while also maintaining full-time tenure-track lines* for truly distinguished new Ph.D.s—say, candidates from Yale or Chicago who've respected their mentors and haven't caused trouble. Yet the only difference between Gilman's proposal and Yale's attempt to eviscerate its local labor unions is this: Gilman thinks *his* proposal will be attractive to administrators, faculty, new Ph.D.s, and undergraduates alike. "Indeed," he writes, "postdoctoral mentored teaching fellowships will provide a real model for undergraduates who may wish to enter graduate school in the humanities" (5). Thankfully, Gilman does not go into detail about what kind of undergraduates would be enthralled at the prospect of attending graduate school for seven to ten years with the hope of eventually becoming a two-year Optionally Mentored Fellow at $20,000.

I have tried, in these pages, to analyze what I regard as the deeply destructive response of Yale faculty to the prospect of graduate student unionization, and I have taken that response as a harbinger of future labor relations in the academic professions. Further, I have tried to link that response to broader tendencies in the leadership of the MLA, an ostensibly "professional" organization that should, if it is going to serve any useful professional function, be defending professional standards for the treatment of its most impecunious and vulnerable members. But I hardly know what to make of my own analysis. When Patricia Meyer Spacks served her term as MLA president in 1994, she addressed the job crisis by candidly admitting that she had "no idea" how to address it (3); Sander Gilman, by contrast, has come forward with a considered, detailed plan for redressing the crisis, and his "plan" turns out, instead, to be a blueprint for dismantling what little job security still exists in academe. I am compelled to conclude that some faculty would do better to ignore the job crisis than to attempt to speak to it, for when they speak to it they sound strikingly like the faculty at Yale: overidentified with the budgetary priorities of university administrations, clueless about their relation to American higher education at large, and all too willing to sustain the profession's ever-dwindling positions of privilege by assigning basic undergraduate instruction to underpaid and overworked adjuncts, "teaching fellows," and graduate students.

By opposing and finally breaking GESO, Yale faculty set an awful precedent for faculty and administrators elsewhere in the country. The MLA, in turn, committed both a tactical and an ethical error by not including GESO spokespersons in their mass mailing to members of the profession with regard to the Yale resolution; and when it comes to professional leadership with regard to the job crisis, nothing could be worse than to have Sander Gilman's postdoc suggestion fall into the hands of cost-conscious administrators. But worst yet—or, perhaps, best of all—Yale faculty and the MLA leadership have now sent an unmistakable message to graduate students, adjuncts, and part-timers everywhere that their nominal spokespersons and their professional organizations are singularly ill suited to represent their interests, and may in fact be best suited, on the contrary, to the desperate, misguided preservation of systems of prestige and reward that are no longer defensible in American higher education's post-Fordist economy. By the AAUP's most recent count, part-time faculty now make up approximately 45 percent of the American professoriate; and at many large American universities, graduate students teach more than half the introductory undergraduate courses in all fields. All told, adjunct faculty and graduate teaching assistants now make up the bulk of the workforce in

U.S. higher education. The time has come for that heretofore silent majority to take matters into its own hands.

NOTES

1. Lafer can be reached at Glafer@aol.com. I have relied on Lafer for much of my understanding of union policy at Yale.

2. Here and throughout this essay, I need to exempt a handful of exemplary individuals, such as Michael Denning, Hazel Carby, David Montgomery, and Rogers Smith, among others, from my wholesale castigations of "Yale faculty." I owe Michael Denning, in particular, a number of accumulated debts in the writing of this essay, since he has been one of my major sources of information on the Yale strike, as well as a keen editor and consultant on the various editorials and brief articles I wrote in January 1996 when it looked as if the *Nation* were going to run a story on the events at Yale.

3. The alacrity and decisiveness of the NLRB ruling were something of a surprise, not only to officials at Yale (who immediately promised to contest the decision) but especially to those of us sympathetic to GESO, who expected the machinery of federal justice to operate about as quickly as the Equal Employment Opportunity Commission under the stewardship of Clarence Thomas. The NLRB took less than a year to rule in favor of GESO, announcing its decision in November 1996 to file suit charging Yale with "illegal retaliation" against graduate students engaged in legitimate organizing activities, and requiring Yale to pay (a) back wages to certain teaching assistants, and (b) fines for breaking federal labor law. The NLRB decision, delicious though it may be, is being fought by attorneys for Yale as this book goes to press. For the record of the NLRB decision, see Greenhouse B6.

4. Brooks, quoted in Eakin 60.

5. Of the disciplinary hearings of early 1996, Cynthia Young writes,

> it was clearly no coincidence that all three of the strikers charged [the third was Nilanjana Dasgupta] were members of the Team Leaders' Committee, GESO's leadership council. However Dean Appelquist insisted that we had been individually identified by our faculty supervisor, because two of the professors involved—Sara Suleri-Goodyear in my case and David Brion Davis in Diana Paton's—wrote letters requesting our grade records and then referred our cases to the Dean when we refused to submit them. (191)

6. No claim is more hotly contested by antiunion faculty than this one. Yale president Richard Levin insists, for instance, that graduate students teach only 3 percent of the courses above the freshman level; but that figure relies on Yale's insistence that teaching assistants are not to be counted as "teachers" for the purposes of calculating figures on "contact hours." However, according to a comprehensive report compiled by Yale graduate students, *True Blue: An Investigation into Teaching at Yale,* graduate teaching assistants in the humanities and social sciences spent 864 hours in the classroom each week, whereas full-time faculty spent 756.5 hours.

3

PROFESSIONAL OBLIGATIONS AND ACADEMIC STANDARDS

Over the past five years I seem to have earned for myself a somewhat schizophrenic—or at least double-voiced—role as an academic cultural critic. On the one hand, I have taken a number of opportunities to defend and explain contemporary theories and practices in the liberal arts and sciences; on the other hand, I have more recently taken a number of opportunities to criticize contemporary academic practices that touch on the structural and economic determinants of the profession of teaching in American higher education. There should be no mystery as to why I began taking more of the latter variety of opportunities *after* being granted tenure in 1993: as I saw it, American academe is in need of defense *and* reform, and if you're going to talk about reform and be taken seriously, you'd do well to make sure you have someplace to stand before you take your stand. In this sense I am a classical liberal reformist, and I hope and expect to be pilloried in precisely these terms: I believe in the possibility of changing the system from inside, and I believe that among the many reasons we should defend the institution of tenure is that it makes it possible for professors to be critics and reformers of the system that houses them.

My own university, for instance, is currently engaged in a costly and foolish legal battle, trying to deny its graduate students the right to unionize—despite the fact that current case law clearly suggests that graduate students at public universities have all the rights of any other public sector employees, *and* despite the fact that the November 1996 NLRB ruling on Yale suggests that graduate students at *private* universities enjoy those rights as well. I consider myself a loyal employee of the University and the State of Illinois, but I do not believe that my tenure at this university entails my agreement to all the initiatives and legal opinions of my administration and Board of Trustees. So, for instance, when in the fall of 1996 I was asked by Illinois's Graduate Employees Organization to speak at a rally, the purpose of which was to protest outside the Board of Trustees' meeting and draw attention to the Trustees' antiunion position, I agreed to march and speak as the GEO's faculty supporter regardless of the fact that my administration had sent all faculty a memo discouraging them from participating in or supporting GEO union drives.

I have to admit that I was personally pleased to oppose my administration and trustees in a cause the rightness of which I am convinced. I even liked delivering an address via bullhorn for the first time in my life (I was only six years old in 1968 and have had to rely on the *New Criterion* for the story of what really happened in the sixties). But at the same time, after a few years of speaking on the subject, formally and informally, at conferences and in the department halls, to students and to colleagues, I cannot continue this aspect of my double-voiced academic life without feeling a good deal of ambivalence. The pain, anger, disappointment, bitterness, and confusion this crisis has caused us—and by "us" I mean the students and recent Ph.D.s most affected by it, as well as those faculty who have made it their concern—sometimes produce in me a kind of survivor's guilt. In such moods I begin to hope that in some future decade if not some future life, I will be so fortunate as to see, and be asked to address, a rather different crisis in the profession: The Job Boom: Are Students and Scholars Being Corrupted by All the Money Flowing into the Humanities?

Likewise, serving three years as Illinois's placement director, at a time when a substantial majority of our candidates were not "placed" in any full-time academic position whatsoever, has often made me want to avoid thinking about the crisis in graduate studies even though I am convinced that over the long term, our current graduate placement rates will do more to damage the profession than a trainload of Lynne Cheneys and Roger Kimballs. Pundits and politicians have begun to call for the closing of graduate programs *in engineering and the applied sciences,* on the grounds that those programs are producing 25 percent more Ph.D.s than the job market can handle (see Greenberg). Surely most humanities professors would exult over something so utopian as a 75 percent placement rate—and surely the humanities are much more vulnerable to fiscal cost cutting and legislative cynicism than are the sciences. Yet most humanities professors seem less concerned about this than about the possibility that their younger colleagues are not paying enough attention to the beautiful and the sublime.

I say this with a certain bitterness, since until recently, media coverage of American universities, with few exceptions, seemed to suggest that the crisis in higher education was that teachers didn't teach enough, or taught the wrong things to the wrong people. Indeed, in the summer of 1995, George Will and Heather MacDonald, those feisty intellectual lapdogs of the Right, accused university writing instructors of poisoning composition courses with a mixture of deconstruction and politically correct feel-good exercises in self-expression. Never mind the fact that university writing instructors, significant numbers of whom are graduate students, are often the only people in higher education who concern themselves with the quality of student writing; Will and MacDonald are not about to blame engineering and commerce schools for their inattention to undergraduate prose. No, what was most remarkably offensive about this latest offensive was Will's conclusion, which proved that conservatives do care about economic inequity after all: "The smugly self-absorbed professoriate that perpetrates all this academic malpractice is often tenured, and always comfortable. The students on the receiving end are always cheated and often unemployable" (A22). It is a curious world Will describes here, where business majors hunger desperately for the ministrations of English teachers, the powerful caste charged with the sole and terrible responsibility of fixing these students' prose and rendering them employable. Makes you want to start a petition calling for term limits on smug, wealthy Beltway pundits, does it not?

Thankfully, a few major newspapers have begun to call attention to the downsizing of university faculty and academe's increasing reliance on adjuncts and part-timers. But I feel safe in saying that the employment patterns in American universities have still not drawn sufficient public notice or outrage. For example, I draw your attention to an article entitled "The Crisis in the Ph.D.," by Cedric Fowler. It tells a story of rising student enrollments in graduate and undergraduate education; rising job requirements for tenure-track faculty and composition instructors; pay cuts; doubling and tripling of class sizes; and most of all, a drastic oversupply of new Ph.D.s:

> Unfortunately the graduate student enrollment has not shrunk in any way comparable to the shrinkage in college jobs. Teachers who have lost their positions and find themselves with a little money turn to further training in the interval of waiting for another appointment. People who have not completed the Ph.D. decide that now is the best time to finish, in the optimistic faith that more training will help them when the colleges get on their feet again. . . . It is more than likely that a further two thousand Ph.D.s will be released on the nation this year.
>
> What will become of most of them nobody knows, for there are almost no college positions available. . . . Over-supply will only grow greater, even after the return of prosperity to higher education, if the present flood continues to pour out of the graduate schools. (41–42)

What's striking about this article, I think, is that it was published in June 1933. That fact alone might lead us to believe, as so many contemporary commentators seem to do, that the employment crisis in American universities is nothing new: 'twas always thus, that Ph.D.s flipped burgers and composition instructors worked for scrip. But I think this conclusion suffers from misplaced emphasis: the point should be that we are justified in referring to the current era of college employment as a Great Depression, and if we remember that the previous Great Depression was resolved only by massive federal action and a global conflagration, then perhaps we can cure ourselves of the wishful thinking that tells us we are merely weathering a downturn in the latest business cycle.

Now that I've established that the prospects are horrible for the profession of literary study and unlikely to improve substantially, what remains to be said? I'll concentrate on two areas, neither of which offers much in the way of a "solution" to what is probably, for new Ph.D.s, an intractable structural problem of underemployment. First, what kind of obligations must faculty and their professional organizations assume in such a crisis? Second, what kind of standards must graduate programs maintain in their curricula, their admissions procedures, and their goals for professional training? In what follows, I don't mean to sound peremptory by saying that such standards and obligations *must* be upheld; every one of my suggestions is open to further challenge and revision. But I want to stress the ethically binding nature of these obligations: however we construe them, they should—I mean, they *must*—be understood as part of the conditions under which departments in the humanities will hereafter operate. For among the things the profession of college teaching needs most urgently at the moment is a working—and, when need be, enforceable—definition of "professionalism."

I'll start by addressing the role of professional organizations. In 1994, when Cary Nelson and I published our brief essay in the *Chronicle of Higher Education*, "Graduate Education Is Losing Its Moral Base," we included in it the suggestion that professional organizations needed to get involved in the process of making difficult recommendations for shrinking—and, in some cases, closing—graduate programs, on the

PROFESSIONAL OBLIGATIONS AND ACADEMIC STANDARDS

grounds that "neither departments nor their own institutions can be counted on to do so" (B3). In the past few years I have repeatedly been told that such language dangerously suggests that legislators and other external meddlers should be given control over graduate education because we academics cannot manage our internal affairs; I have also been told that Cary and I are unreasonably asking the MLA to police the size and viability of graduate programs. But since the MLA already gathers a great deal of information on the academic job market and the size of graduate programs, I do not see why it cannot make recommendations based on its findings. As Stephen Watt has pointed out, we cannot expect (and should not want) the MLA to maintain the kind of extraordinary professional control over wages and working conditions that the AMA has achieved over the supply of physicians in the United States (Watt 33). Nevertheless, it is possible to say that the MLA can be more aggressive in the broader sense of advancing the interests of the profession and in the narrower sense of censuring those departments that ignore the MLA's standards for job searches.

I want to distinguish this claim from the claim that the MLA itself is a major player in the job crisis. There is no point, I believe, in attributing blame to the MLA for not immediately ameliorating conditions it did not bring about. Be that as it may, I have—I should say Cary Nelson and I have, since these are part of our ongoing conversation—two small, practical suggestions for the near future. Both, we think, would enhance the MLA's image within and outside its membership, and both would make the most of the MLA's already formidable information-gathering apparatus.

First, the MLA should try to ascertain, as closely as possible, the number of job seekers each year. Currently we have a system that tells us how many jobs are advertised annually in the MLA Job List, but no reliable figures for the "supply" end of the equation. When, in 1994, the MLA released statistics that indicated that 51.1 percent of new Ph.D.s had found full-time, tenure-track employment in 1991–92, it wound up publishing a figure that no one in the profession could believe; and the reason for this is that the MLA only took stock of job seekers who had received their Ph.D.s in that year (see Huber). No mention was made of

PROFESSIONAL OBLIGATIONS AND ACADEMIC STANDARDS

job candidates who had earned their Ph.D.s in earlier years—and in the light of the past few years, it is easy to imagine that the largest and most desperate group of job candidates would be composed of people who had been on the market more than once, and who are now either unemployed or underemployed in part-time adjunct positions. In future years, as this pool of Ph.D.s swells (partly because of the decline in jobs, partly because of the dramatic increase in graduate enrollments between 1985 and 1990), it will become all the more imperative for the MLA to have a more reliable estimate of how many Ph.D.s are actually searching for jobs. All that would be required is for the MLA to include, in each Job List, a self-addressed postcard; students with photocopied lists could submit a facsimile card, announcing their intention to attend the convention and secure academic employment. I can think of no cheaper and easier way of determining that most important number of all, the annual number of job candidates. The complaint that such a system is open to "fraud" seems to me too bizarre to entertain.

The second suggestion, also, I owe in part to Cary Nelson: the MLA could appoint a standing committee, not unlike those operated by the AAUP, to investigate deceptive hiring practices—particularly where schools are in violation of already existing MLA recommendations. I regard this as less a matter of "policing" than of protecting the professional interests of the membership. As for the argument that the MLA has and can have no "enforcement arm" in such matters, my response is that it does not need one. All the AAUP has is a list of censured colleges, places like Appomattox State Bible Seminary and Whiskey-a-Go-Go University of the Sierra Nevada where faculty are hired on a week-by-week basis or are forced to double as bartenders for administration functions. But the AAUP is less concerned with changing practices at such marginal schools than with ensuring professional treatment of faculty at larger places—Harvard, Indiana, Grinnell—that actually do most of the hiring. The idea is that Harvard, Indiana, and Grinnell can be induced to play fair by the threat of having their names added to the list of odd miscreants like Whiskey-a-Go-Go, which most administrators consider fairly bad publicity. And although this is a more "symbolic" gesture, the MLA could in the future refuse to run announcements of job openings from "censured" schools. The counterargument I have heard from MLA staff is that such schools would merely advertise their jobs elsewhere, but I do not see this as a barrier to taking action that would alert potential candidates to the fact that the school has not met the profession's criteria for job searches.

I said I had two suggestions for MLA action in the future. I have also a third suggestion, but it is so unlikely to have any impact that I might as well not mention it. Still, here goes: move the convention to March. I have now heard every argument against moving the convention from December—academic schedules, hotel rates, timing for interviews—but I have also seen an alarming growth in the number of academic jobs advertised in October but not funded in February. More than once I have had to console students whose only MLA interview was with a department whose search was eventually shut down in the spring for lack of funding; nothing makes those students, or me, or, for that matter, a school's search committee, feel more helpless and futile. Having the annual convention *after* most schools receive their funding allocations for the year will not prevent this from happening altogether; it certainly will not prevent search committees from working in the fall to fill a position that does not exist in the spring. But perhaps it will help reduce the number of times this happens, and in times like these that is no small consideration.

It has long been maintained that late December is the only time when all members can meet, since the quarter system and the semester system are otherwise incompatible. However, rethinking the MLA convention is not merely a matter of fiddling with dates. Why should the convention take place over four days at year's end rather than over a long weekend in the spring? Why, for that matter, should it showcase (if that is the word) the delivery of nearly two thousand fifteen-minute papers? The convention could be an occasion for focused reflection on the state of the profession; as currently conducted it is little more than fodder for ravenous journalists looking for a quick laugh at the expense of teachers in the modern languages. Of course, the convention's chief *structural* aim is to provide a forum for job interviews—which means that the convention is largely kept afloat by job applicants. And the MLA does well to structure its registration fees accordingly, charging graduate students less than half the fee paid by professors. But why should we collect any but a nominal ten-dollar fee from job seekers who do not now have full-time tenure-track jobs, so long as they are attending the MLA for job placement? Why should the convention not be held in smaller, less expensive locations? Why should it not be reorganized to serve the interests of interviewees and interviewers above all?

This is only one small area in which we need to recalibrate our professional priorities. If we want our profession to survive the depression more or less intact, we will have to rethink more broadly the entire range of college teachers' obligations to each other and to their student-apprentices. This is the hard part of professional housecleaning; it involves not only a discussion of the invisible work faculty do on graduate, departmental, and campus committees, but a discussion of the politics of curricular design and early retirement as well. Let me deal with the matter of early retirement first, since that has so far proven to be the most incendiary and the least understood of the suggestions Cary and I have made. In our *Chronicle* essay and in *Higher Education under Fire*, Cary and I wrote that "institutions should devise legally sound early-retirement packages for those faculty members who are neither effective teachers nor productive scholars. . . . For we need to confront the fact that we are driving talented new teachers and scholars out of the profession while retaining some incompetent faculty members with tenure" (B2). Though we nested this suggestion in among six or seven more innocuous recommendations, it has not failed to catch the eye of most of our readers; within days after publication of that issue of the *Chronicle*, no less, one of Illinois's most distinguished senior scholars approached me and asked me, tongue firmly in cheek, where faculty over fifty years of age could turn in their badges and resign. (I replied that I could not answer such a question myself and would have to turn it over for review by the Star Chamber.) Oddly, however, some of the most hostile responses we've received to this proposal have come from graduate students, one of whom charged in a letter to the *Chronicle* that we were trying to weed out the field of job seekers and job holders so as to pave the way for more "superstar" faculty. More recently, Joseph Aimone, the vice president of the MLA's graduate student caucus, has written in the NCTE's *Council Chronicle* that "limiting early retirement to faculty who are neither effective teachers nor productive scholars misses the point." Aimone goes on to say that we should induce *effective* faculty to retire, because the "deadwood" standard is a dangerous one:

> Effective teachers and productive scholars need to be induced to retire. Retirement for academics only means that they teach when they want to rather than when the whistle blows and write without care about tenure. . . .
> . . . Allowing a standard that looks for "deadwood" is inviting people with no interest in higher education to make the decisions about who should and should not work. Voluntary, systematic retirement of senior professionals, combined with a vigorous effort to ensure hiring tenured people behind them, would make a difference. Is this likely? Ask Nelson and Bérubé—would they retire in midcareer for the good of the profession? (5)

I'm not sure what fuels the antagonism of this response, though I suppose I have a clue. But without announcing my own early retirement date, let me address the question of how we propose to carry out early retirement policies as fairly and as painlessly as possible.

The issue for me is not that we should look for "deadwood"; the issue is that we should determine the professional obligations and expectations for all tenured faculty, and then, if certain faculty members are egregiously flouting those obligations, then we ask that they retire to allow more capable and committed people to teach in our place. There are two relevant questions here, and they are both, at bottom, questions of professional ethics. One has to do with the standards to which we hold our senior faculty and our job applicants; the other has to do with how the faculty carries out its own departmental functions.

The job requirements for faculty are notoriously loose—so much so that one enterprising professor recently managed to teach at two universities simultaneously, without attracting notice, for a number of months before he was found out and fired from both. Some of us serve on an array of departmental committees, dissertation committees, campus committees, editorial boards, and internal review panels; others of us teach their classes and go home. Most of us are working sixty-hour weeks and a small handful of us are working six. Now, I realize how dangerous it is to admit this at a time when most legislators and journalists are all too eager to characterize *all* faculty as tenured, comfortable freeloaders. And I realize that every organization inside and outside academe has these classes of people: workers who foolishly demonstrate their efficiency and competence so often that they are assigned to every important task, and workers to whom meaningful authority is never delegated and by whom it is never desired. Academe, however, has arguably too high a tolerance for the latter group; and when that latter group contains full professors, as it sometimes does, you can wind up with a system that not merely tolerates but *rewards* workers for performing substantially less work than their colleagues. I'm sure I'm not revealing any family secrets in saying this. But it has often seemed to me that academic departments can work very much like dysfunctional families where the parents clean their teenagers' rooms because it's not worth the trouble of asking the kids to do the job themselves. No one asks certain people to contribute to the maintenance work of the department in service and advising, no one talks about *why* those people are never asked to contribute, and the workload imbalance continues by mutual consent even as it worsens.

I'm stressing faculty "service" for a number of reasons. First, I have— and had—no intention of restricting the professoriate to people who want to be one-person publishing houses; that was not what Cary and I meant when we spoke of "unproductive" faculty. Second, I know that effective teaching is very hard to measure, and ineffective teaching equally difficult to spot. And last, because I want as value-neutral a criterion of "professional obligations" as I can get. Here's why: from the perspective of moderately talented faculty who were hired in 1964 when they had their pick of jobs to choose from, the situation I've described so far can look rather different. Why should we retire, these people may say, and let ourselves be mowed down by a horde of Stalinist feminazi ideologues who want to destroy literary study as we know it? For it can't be denied that the profession practiced by senior faculty and the profession envisioned by junior faculty and Ph.D. candidates are often two radically different things—and as I'll explain in more detail in the next chapter, each generation is thereby tempted to construe the other as the source of all the profession's ills.

The problem here is that if we define "incompetence" to cover faculty who have lost all intellectual interest in the state of their discipline, we let ourselves in for more nastiness than anyone can imagine. One professor refuses even to discuss feminist interpretations of Shakespeare, and dismisses the student who brought up the question; another is unaware that the anthology he's using now includes women writers, none of whom he's assigned to his students. Are these people incompetent? Are they irresponsible? Many of my colleagues would say so. And yet you certainly can't try to retire someone on the grounds that he hasn't kept up with the field, especially when the teacher in question believes that the field basically died in 1970 and has been churning out twaddle ever since. Promoting that kind of criterion for faculty competence, therefore, seems to me to invite a host of intractable political problems, not the least of which is that it violates faculty members' academic freedom to believe that their refusal to keep up with the field, or to discuss feminist readings, is a sign of intestinal fortitude.

So when I bring up the unpleasant topic of early retirement, I don't mean to say that we should consider early retirement policies for people who don't write books every three years or perform in the classroom like Robin Williams in *Dead Poets Society.* I'm also not saying that we shouldn't trust anyone over fifty. All I'm saying is that many academic departments house one or two faculty members who simply do not meet their professional obligations as scholars, teachers, and colleagues. And in a job crisis as severe as this, it is unconscionable to allow those few people to work their six-hour jobs while overqualified applicants serve their time as adjuncts and freeway fliers. Let me add this: if we do not address this problem—if we do not even admit that there is such a problem, however slight it may seem—we will soon see the dismantling of tenure. We are already witnessing the *de facto* eradication of tenure in the wholesale conversion of full-time positions to adjunct positions; but I fear we may also see the *de jure* elimination of tenure as schools implement more "austerity" policies in hiring and legislators in Texas and elsewhere demand post-tenure review and punitive or bureaucratic forms of "accountability." And if tenure is wrongly construed as a protection not of academic freedom but of manifestly incompetent and irresponsible workers the like of whom no other profession would tolerate, then its days are numbered, and the profession will be subject to even more brutal conditions wherein only the wealthiest universities will bother to hire full-time faculty. The best way to preserve tenure, it seems to me, is to create meaningful forms of internal faculty review—and to create them ourselves, rather than waiting to carry out the dictates of state legislators.

I cannot expect to win anything like universal consent for this proposal. It is all too possible that "post-tenure review" will become, for those universities that implement it, a means not of improving the quality of faculty but of dismantling whatever job security remains to the professoriate. I have had to acknowledge the danger of my position on a number of occasions, none more dramatic than the time I spoke on this subject in February 1997 to the Faculty Senate of the City University of New York—an institution that has seen, in the past twenty years, its number of full-time faculty drop from 15,000 to 5,500. But it is crucial in the face of numbers like those, I think, to stress that I am not talking about *firing* faculty—or even about finding rationales for downsizing the professoriate still further. I am talking about early retirement agreements that might help ease the job crisis: this is not an austerity program for academe, but a call for replenishment of the ranks. My proposal, moreover, is meant not to take decision-making power out of the hands of the people most concerned with education, but rather to *give* those people—in this case, department chairs—that kind of decision-making power. Indeed, constructing a professional standard that measures faculty by whether they responsibly meet their professional obligations is, I believe, the only way that faculty can retain the power to determine the composition of their departments.

I should repeat as often as possible that very few faculty actually fail to live up to most of their professional obligations, and that the point of my proposal is not only to prune departments of genuine deadwood but also to devise a system, a code of ethics, whereby all faculty know what their responsibilities are. (Stephen Watt, for instance, informs me that the English department at Indiana has a clause in its bylaws that stipulates that the supervision of a dissertation is part of one's professional teaching load, and should be considered the equivalent of teaching a course. The clause is toothless, of course, but remarkable and enviable nonetheless.) But as long as we're concerned with numbers, let's imagine for a surreal moment that there is only one thoroughly unprofessional faculty member for every five English departments, and let's say we collectively negotiated their early retirement next year. The MLA *Job Information List* would grow by 700 positions. When you recall that the October 1992 *JIL* listed 620 positions and the October 1993 *JIL* listed 624, you'll probably reach the same conclusion I've reached: early retirement packages are the single most powerful means we have for alleviating the current academic depression. The challenge is to contrive legally binding methods for tying one issue to the other, lest a university administration be tempted to replace forty retirees with forty part-time adjuncts. Early retirement packages, in order to be ethically defensible, absolutely must be predicated on the hiring of full-time, tenure-track faculty. Anything less will simply accelerate the dismantling of the profession.

I should also stress that all of us who fondly imagine that we will be passed over by such policies will have to make some sacrifices as well. Many faculty in humanities departments do not make enough money to retire, and if early retirement packages are to work for such people, they might need to be awarded substantial pay raises for a number of years prior to their retirement. Cary Nelson has suggested in the November-December 1995 issue of *Academe* that a humane guideline for such packages should state that no retiring senior faculty member with thirty years of service will be allowed to retire on a fixed annual income smaller than that of a first-year assistant professor in the arts or humanities. As Nelson writes, such a policy

> states the problem both baldly and realistically and sets an individual retirement package goal that few are likely to regard as a reward for incompetence. It also acknowledges the real financial

risk some underpaid faculty members face at retirement time, while asserting that universities have no business trying to sustain higher disciplinary salaries after retirement for those faculty who have not performed competently. There is no reason why a retired marginal commerce professor should earn more than a retired marginal philosopher. (25)

This proposal is no idle suggestion: imagine a humanities professor hired in 1970 at a salary of $10,000 who made $22,000 in 1985 and makes $35,000 today. To retire at a decent *entry-level* salary, that professor might require salary increases of 50 percent over a few years. At the very least, that means that everyone else in the department may have to forgo their own piddling raises, or even endure small cuts, so that the department will be able to make new hires in the near future. And the point is not to bar the retiree from the classroom; on the contrary, emeriti should be encouraged to teach a class whenever and however they want. They will simply be officially relieved of the myriad professional obligations they are no longer fulfilling.

Now for unpleasant draconian suggestion number two: we will abrogate one of our chief responsibilities as teachers if we do not discourage more students from pursuing the Ph.D. In the past, I have heard this proposal denounced as "elitist," as if I were suggesting that everyone close their graduate programs save for the Ivies and Berkeley. On the contrary, I am glad that our profession gives less weight to the prestige of the degree-granting institution than does law or medicine, since high-quality Ph.D.s can literally come from anywhere. With that in mind, I suggest that *all* graduate programs, whatever their rank in the collective professional imaginary, decrease the number of students they admit to the Ph.D.

I noted above that I anticipate criticism of my early retirement suggestions; I have already gotten many earfuls of criticism with regard to my position on the size of graduate programs—most of it, interestingly, from my colleagues on the academic Left. The harshest response Cary Nelson and I have received has come from a pair of Marxists, Jim Neilson and Gregory Meyerson, who, in a 1996 issue of the *minnesota review,* painted us as corporate stooges in progressives' clothing:

Most troubling are their suggestions to limit graduate admissions and to close marginal programs. If this proposal had been made by a right-wing politician about undergraduate education—that is, if in order to address the problem of unemployed university graduates it was proposed that fewer students be admitted into college and that marginal schools (i.e., community colleges, poor private institutions, and inferior branches of state universities) be closed—this politician would rightly be accused of denying educational opportunity to poor, minority, and non-traditional students and of increasing elite advantage. But since it's been made by academic leftists, this proposal has been greeted as a pragmatic attempt to lessen the exploitation of graduate students. (270)

Neilson and Meyerson go on, in a footnote, to claim that "although we don't mean to impugn their motives, it's worth noting that this proposal requires no sacrifice of Bérubé and Nelson." Indeed, they claim, Nelson and I stand to profit by our proposal: "with fewer graduate programs, their status (both in professional reputation and financial reward) is likely to be enhanced" (272). That's a nice touch, coming from people who don't mean to impugn my motives. But it doesn't make sense. Under the current dispensation of rank and reward, Illinois does not become more prestigious when North Dakota or Valparaiso closes a program, and Illinois is not so lofty a location as to be exempt from the sacrifices attendant on an austerity economy. As a matter of fact, Illinois's English department has drastically cut back on the number of graduate students it admits (while maintaining a diverse student population that gives the lie to the "younger, whiter, wealthier" line); at around one hundred, ours is the smallest graduate program in the Big Ten—half the size of places like Indiana or Michigan, and less than one-third of the size of the Virginia program from which I graduated in 1989. Most of our recent Ph.D.s, for their part, are quite bitter about their lousy job prospects, and some have emphatically suggested closing the program altogether. Neither Cary Nelson nor I have "benefited" from this development, and we sure as hell haven't pocketed any spare change by it, either.

But Neilson and Meyerson's critique involves some crucial switching of the dice—in this case, a shaky and mistaken analogy between undergraduate and graduate education. Nelson and I have actually argued for *expanded* access to the B.A. and M.A. degrees, but when it comes to large Ph.D. programs in the humanities, we advocate more central planning in place of the laissez-faire, let-the-student-beware economy we have now. We believe in expanding the educational franchise, but we don't believe in expanding cheap labor pools. *That,* of course, is why the Bérubé/Nelson proposals have been understood as a means of opposing the economic exploitation of graduate students; it's certainly not because Nelson and I have been mollycoddled by the liberal media.

Neilson and Meyerson have one more important criticism of my proposal to shrink doctoral programs, and this criticism, I believe, speaks for itself:

> Especially in recent years a graduate education in the humanities may equally be a political education, a means by which students learn to read the historical, social, and economic truths hidden and distorted by capitalist culture. Bérubé and Nelson's proposal ignores this important justification for maintaining the wide availability of graduate studies in the humanities. Bérubé and Nelson propose to reduce enrollments in graduate programs, making the demystifying and consciousness-raising potential of these programs available to a privileged few; ironically, their solution to the crisis in higher education is, in effect, to *limit* public access. (271)

I hope it will be clear why I did not anticipate this argument when I first began to think about whether the country needed dozens of graduate programs in English with two hundred or more students. At the moment, English departments may be placing fewer than 20 percent of their Ph.D.s in tenure-track jobs; the other 80-plus percent are unemployed or employed in temporary positions at starvation wages without benefits, and because they're on the job market year after year, the MLA job placement statistics don't count them as job seekers. But in this rotten economy, according to Neilson and Meyerson, we should "maintain the wide availability of graduate studies in the humanities": we should admit students to programs of study in which they will devote seven to ten years of their lives, during which they will teach introductory undergraduate courses at about $2,500 per course (without benefits), so that, after a decade, they can have about an 80 percent chance of teaching piece-work courses at local colleges for about $1,250 per course (without benefits). And the reason we should continue, and even expand, this organization of pedagogical labor is that *it will teach graduate students about capitalism.*

For my part, I fear that the Neilson/Meyerson plan will work only too well. Graduate students will learn about capitalism, all right—not by having their collective consciousness raised in the Marxist graduate seminar, but by working in academe's salt mines until middle age or there-abouts, whereupon they will find they are the owners of a postgraduate degree that is practically useless.

However much it may outrage one particularly befuddled wing of the academic Left, then, my rationale for cutting doctoral programs is both simple and reasonable: the M.A. degree does not offer students the kind of long-term exploitation, disappointment, and highly specialized training now associated with the Ph.D. It is one thing to put in anywhere from one to three years in postgraduate study, and then pursue a career in publishing, journalism, advertising, law, or teaching secondary school; it is another thing to devote a decade or more to professional training for a profession in which there are no jobs. Not only is it easier to change career directions at age twenty-five than at thirty-five or forty; it is also easier to apply for jobs outside the modern languages with an M.A. than with a Ph.D. It should go without saying that it is also easier to pay off two years' worth of loans than ten, or endure two years of starvation wages rather than ten.

The problem with this advice is that it is so hard to give. I have not yet met the student who could contemplate being told to stop after the M.A. without hearing the suggestion as a wholesale rejection of his or her very person. Faculty can try to say, "you had best not pursue this—it's highly unlikely to result in anything like a rewarding career," but few students will hear this as anything but "you had best not pursue this—you're not smart enough to make it." And since faculty and graduate students tend to have a tremendous emotional investment in having their intellectual talents validated by people they respect, it is often impossible to hear a discouraging word about one's prospects in the profession without taking it as a judgment on your net worth as a conscious being. Add to this the fact that at many schools, the M.A. degree has no function whatsoever save as a qualification for the Ph.D., and you have a system that strongly discourages faculty from giving students honest advice. I am not asking faculty to shout "fire" in a packed theater so as to thin out the crowd; rather, I am asking faculty to tell students not to enter burning buildings—and I am asking graduate students to take the advice in good faith.

But if our goal is to restrict access to the Ph.D. while maintaining or expanding access to the M.A., then we have some problems to face. First of all, it is not clear what the M.A. means; in some states it counts toward certification for high school teaching, in some states it does not. In some states it exists as an incentive for current high school teachers to get extra professional training in their disciplines, and, maybe, a raise. Second, we run the risk of making the first years of graduate school even more tense than they already are, by fostering a competition for slots to the doctoral program; schools that do not already work this way may well be loath to try it. Third, we run the risk of creating a shortage in the cheap labor pool every sizable university needs to survive: as Cary Nelson writes in a recent issue of *Social Text,* it would cost our own English department over $4.5 million to staff all our courses with teachers who were paid the average department salary for faculty, $3.5 million to replace all graduate students with entry-level assistant professors (131). As Illinois's fledgling Graduate Employees' Organization rightly says, the university works because *they* do. And last but not least, we threaten the very existence of M.A. programs who hope to send their best students to doctoral programs elsewhere.

All these problems are then compounded by the fact that any responsible teacher who cares about the state of literary and cultural study in the United States is inevitably a personal advertisement for the discipline: we believe that this is among the most intellectually challenging fields in the contemporary college curriculum; we think the kind of intellectual work we do is fundamental to what it means, socially, politically, and psychologically, to be human; and we therefore think this kind of study is

valuable even if the vast majority of our fellow Americans do not agree—
and even if we don't believe there is such a thing as "intrinsic" value.
There is a sense in which every good teacher *wants* to recruit his or her
most promising students for the field; simply think for a moment of
what it would be like if we asked our colleagues to behave in the
classroom as if they thought literary study were a vain and pointless
exercise.

So we need graduate instructors, but the apprentice system that justi-
fies their status as cheap labor is in deep crisis; we promote the extraordinary elasticity and interdisciplinarity of intellectual work in the humanities, but we need to discourage many of our brightest students from
hoping to join the field; and the only plausible solution to this impasse
lies in redefining an M.A. degree that at the moment has almost no
definition at all.

Our hope, I hope, lies in strengthening the ties between the M.A. and
high school teaching—not only in the sense that we should try to offer the
degree to more high school English teachers, but also in the sense that we
should try to imagine the teaching of high school English as a worthy and
appropriate career for midlevel graduate students. In my experience, suggesting to students that they might teach in secondary schools has been a
little like nominating one's colleagues for early retirement: *here's the M.A.,*
students hear, *and here's the map out of town. Don't bother thinking about
the power and prestige of being a college professor—here's your free pass to
Central High.* To students who regard high school teaching as something
unspeakably worse than college teaching, I have shown the following job
announcements, all of which, happily enough, appeared in the same issue
of our departmental newsletter one fine spring day:

> College of Lake County. Tenure track position to teach English
> composition and literature. Course load is 5 sections per semes-
> ter. Position begins fall 1995.
> Gustavus Adolphus College. Seeking a person to teach from
> September to February as a replacement for a professor on leave.
> The individual will teach three courses (Creative Writing and
> Ethnic American Literature) in the fall (Sept. to Dec.) and one
> course in the January term.

Lincoln Land Community College. Full-time tenure-track position to teach five classes. Should have experience teaching the writing process at different levels to both traditional and non-traditional students and should be able to teach the full range of the lower division curriculum.

These cheery notices were soon bested by another local school, which advertised a position that would carry tenure without promotion for a salary in the low twenties: to the lucky candidate, a lifetime instructorship without hope of further professional reward. I show students these notices not merely to frighten and depress them (though this works like a charm), but also to make a more important point, namely, that some opportunities in high school teaching can offer greater professional autonomy, more substantial intellectual rewards, and better pay than teaching at the college level. As Alison T. Smith, a 1994 Ph.D., writes in the 1996 issue of *Profession,* secondary school teaching is "still an ignored market": even though her own experience teaching high school "proved one of the most rewarding I ever had," still, her colleagues warned her "not to stay there too long lest I be labeled a high school teacher, which would forever destroy my prospects of getting a serious job at the college level" (69–70). Smith now teaches at the Hill Center in Durham, North Carolina, a high school for students with identified learning disabilities, and reports that "the salary, benefits, and level of respect I receive from colleagues are better than what I found at the university level" (72).

Alison Smith's experience, as former Illinois graduate students can attest, is not unique. Indeed, for every student who resents the advice to seek a high school job, I wager, there are ten more who wish they'd heard that kind of advice six or seven years ago. Nor is this strategy merely a matter of cutting our losses; on the contrary, it could be a strategy for dramatically expanding our potential public constituency. The profession as it now operates seems much more interested in producing volume after volume of criticism and theory for faculty and graduate students than in disseminating some of that criticism and theory to undergraduates and high school students. Well-trained, unembittered, comparatively unexploited M.A.s might perform a crucial function in serving as liaisons between graduate programs and public and private secondary education.

Finally, there is the question of how to conceive graduate study itself: how should we design programs of professional training in a profession with no self-description and very few job openings? Should we try to teach to the market, whether that means training students in mastering the details of the counterhegemonic post-excremental sublime, or training them to teach writing across the curriculum? Even if we knew what the market would be like in the year 2005—a surplus of Victorianists! Postmodernists down 1⅛!—would we be justified in redesigning graduate study so as best to allow the tail to wag the dog? Some of my students and colleagues look around and say, what we need here, if we are to survive into the twenty-first century, is queer theory and cultural studies. Others propose a greater emphasis on the traditional periods of literary history, on the grounds that new Ph.D.s will likely have to be broad generalists. Still others recommend that students carry at least a subspecialty in rhetoric and composition.

I did not save this question for last because I find it easiest to answer. I want to argue that graduate programs should recognize that theory is now an integral part of the regime of professional training, so much so that search committees no longer have to stipulate that they want an Americanist who does theory: the theory part, it could be said, goes without saying. And yet relatively few graduate programs offer broad introductory courses in interpretive theory. All too often, the result is that students encounter specialized graduate-level courses in theory without having taken the equivalent of survey courses in theory; "when," my students have asked me, "was I supposed to have read Lacan? Was I absent the day they assigned Bakhtin and everybody moved on to *In the Wake of Bakhtin*?" In my own department, then, the pattern is that theory-laden courses are most plentiful at the opposite ends of the curriculum: at the 400 level, where they are taught exclusively to graduate students, and at the 100 level, where they are taught exclusively *by* graduate students. We have not yet devised an intermediary program by which we would determine what it is that we expect our graduate students to know before they develop their own courses, and in this respect, I will hazard, we are not alone.

To put this point another way, if we conceive of cultural studies, for instance, as something that can be taught only by specialists to specialists, then cultural studies will become just another item in the theory-survey curriculum alongside deconstruction and new historicism, structuralism and psychoanalysis. But if we conceive of cultural studies—and theory more generally—as something that is potentially as relevant to freshman writing as it is to graduate seminars, then perhaps we can begin to make productive use of the multiple theoretical paradigms currently operating in the profession without overspecializing or underpreparing those graduate students who do choose to seek the Ph.D. We can, in other words, escape the illogic of the current system that asks job candidates to be brilliant, original researchers up until they receive an MLA interview, and then to be all-purpose generalists who can teach writing, Shakespeare, and the History of the English Language once they arrive on campus. But we can do this only if we recognize that Ph.D. programs are designed for specific professional training. This means that we need to stop behaving as if advanced graduate study is self-justified, as the pursuit of knowledge and the enhancement of the life of the mind, and that we need to plan to train students for a profession in which theoretical specialists are most often marketable only if they can present the principles of advanced literary and cultural theory in the full range of the lower division curriculum.

In the meantime, one pedestrian point about unions, wages, and benefits. I have already admitted that in most universities, it is impossible for English departments—or most other academic departments—to function without a large pool of exploited student labor. Regardless of what happens to the job market, schools will never be willing or able to pay graduate teaching assistants a living wage for their work. Cary Nelson and I have floated the less expensive but equally quixotic proposal that they be paid enough to live on through the summer, but on the whole, we realize that no one is going to pump a few million dollars into English departments merely because it might make for tolerable working conditions among the teachers who do the bulk of writing instruction for incoming undergraduates. Many faculty, accordingly, do not see the value of graduate student unions, and cannot be persuaded to support their formation, particularly at schools where the faculty themselves are averse to unions and therefore have no capacity for collective bargaining. Indeed, it would even be possible to imagine a critical mass of antiunion faculty at an Ivy League university. But except in such cases of deliberate malice, most faculty are uninvolved in graduate students' efforts to unionize because they do not see the point: graduate students have been eloquent and threadbare since the days of Raskolnikov. 'Twas always thus.

But few older faculty, I wager, have any idea what it is like to live for the better part of a decade without adequate health insurance. Cary Nelson and I have argued that graduate teaching assistants with five years of service should be vested in their universities' retirement plans; we would like to see universities develop sane child care policies as well, but since so few have done so even for faculty, we despair of seeing the day when graduate students are entitled to university child care as well. Health insurance, however, is very much a negotiable item, particularly since most graduate students are relatively young and most universities buy insurance packages for large groups. The faculty who received their Ph.D.s back when health care and carfare cost thirty-five cents cannot be expected to understand the plight of students paying $1,200 for swiss-cheese coverage, but they must be made to. Time and again I have shocked my colleagues by informing them that after I was no longer covered under my parents' dental plan, I waited four years before seeing a dentist in 1985, right before my wedding, and then waited another five until 1990, when I was hired at Illinois. Or by informing them that Janet and I were so thrilled to find in 1986 that our student policy would cover the cost of a hospital maternity room that we neglected to realize that the insurance company would pay only for *Janet's* stay in the hospital—which was no great burden for our insurers, as it turned out, since they billed the room to our newborn baby Nicholas instead. Too late, we realized that if we had given birth to a nonbillable entity such as a cat, we would have been in the clear; but because we had a *child,* we were therefore liable for as much of our hospital costs as could be assigned to a human being.

It took us five years to retire that debt, but that's not the point. Part of the point is that just before Nick was born, I agreed to take Virginia's M.A. exam so that I would qualify as a Virginia alumnus and become eligible for special rates on life insurance available through the Virginia Alumni Association. That, for me, was the function of the M.A. in English: it gave me membership in a group that could buy life insurance. But the larger point is that the profession of English is an increasingly dehumanizing and dispiriting affair for even (or especially) the most ardent lovers of literature and literary study. If the unionization of our graduate students can make our collective enterprise less dehumanizing—and it can—then we are bound to support it. For the still larger point is that a profession that tolerates and perpetuates such conditions is neither professional nor defensible. If you add the working conditions of graduate students to the figure I've cited earlier—namely, that 45 percent of the professoriate consists of part-time laborers—you'll realize that in structural terms, the job of teaching college, for the majority of recent Ph.D.s trying to get that job, has more in common with the job of picking oranges than with the job of practicing law. Whatever our academic standards—be we new historicists or new critics, multiculturalists or Eurocentrists, young turks or old guard—it is our professional obligation to do whatever we can to change that for the better.

4

PEER PRESSURE

POLITICAL TENSIONS
IN THE BEAR MARKET

Nineteen ninety-four was a momentous year for many reasons, and perhaps one of the least important of these was the reemergence, in late fall, of the figure of the Arbiter of Aesthetic Value. Although Rwandans and Republicans made infinitely more important news during the year, something ought to be said nonetheless about how wondrous a moment it is in American culture when a mere literary critic can write a book insisting that (alas) literature has no socially redeeming value, no "utility" at all, and get glowing reviews from practically every major English-language newspaper on either side of the North Atlantic. The idea of "the aesthetic" may be under assault in the groves of academe, but in the general culture, it seems, arguing for the autonomy of the aesthetic and against the School of Resentment can net you a best-seller and an advance of over half a million dollars—*if* you've got the right marketing strategy. The moral is clear: whatever "aesthetic value" may be, it's definitely convertible into a big pile of cash.

Part of Harold Bloom's success with *The Western Canon* stems not merely from Bloom's status *as* Bloom, and not merely from his considerable talent at coming up with sound bites as "provocative" as anything Stanley Fish, Andrew Ross, or Bloom-protégé Camille Paglia can offer the media, but also from the sheer certainty of his determinations of value. Here, at last, is someone who knows what great literature is, and is not afraid to say so. It doesn't matter whether anyone *agrees* with Bloom's determinations; no one human can. The point is that he has made them—like E. D. Hirsch, in a handy back-of-the-book compilation—and thus afforded every literate person an endless supply of what Northrop Frye once called "the literary chit-chat which makes the reputations of poets boom and crash in an imaginary stock exchange" (18). One might say that Harold Bloom has brought to the discussion of American letters some of the thrill of the NCAA basketball and football seasons: could Penn State have beaten Nebraska in '94? was James Jones really faster than James Baldwin in the forty-yard dash? is the Big East (Whitman, Dickinson, Williams) a better poetry conference than the Mississippi Valley (Eliot, Ransom, Tate)?

These are the central questions for criticism, the determinations of value that separate the men from the boys, the thrill of victory from the *agon* of defeat. Indeed, 1994 saw something of a resurgence of criticism in the heroic mode, as Bloom was fortuitously joined by a host of values-hawkers ready and willing to tell us the difference between a four-star poet and a C+ movie. William Henry III, the late drama critic for *Time,* made an impassioned case that some things are just *better* than others, dammit, and so too are cultures that go to the moon superior to cultures that push bones through their noses (14). "I find the blues irresistible," wrote Henry; "I have always liked square dancing. But these are lesser forms of art than, say, oil painting and opera and ballet, because the techniques are less arduous and less demanding of long learning, the underlying symbolic language is less complicated, the range of expression is less profound, and the worship of beauty is muddied by the lower aims of community fellowship" (176). For similar reasons, Henry called for the closing of community colleges (165) and—a nice touch in a book titled *In Defense of Elitism*—"massive layoffs of faculty" (166). The redoubtable Thomas Sowell put the screws to cultural relativism as well, writing that some cultures are "*better* in some respects than others" (4) just as "Arabic numerals are not merely *different* from Roman numerals; they are *superior* to Roman numerals" (5)—a principle few would contest, unless it were stripped of its qualifying clause, "in some respects," and extrapolated to cover moon shots and nose bones. And they were joined by a whole host of Cal Thomases, Bill Bennetts, Dan Quayles, and Rush Limbaughs, wise and learned people who know what virtue is and why they have it and we don't. As a result, the tide of liberal PC relativism was finally turned, and America's quota queens and welfare cheats were hurled from their positions of national power in a stunning electoral landslide some commentators have rightly called the Year of the Man. Perhaps the Book of the Year in the Year of the Man was Dan Quayle's own *Standing Firm,* in which he wonders why Anita Hill doesn't teach at a better school than the University of Oklahoma— since, as a Yale grad and a black woman, "she presumably could write her own ticket" (271). Hill, in other words, must be truly mediocre not to have jumped on the black woman's Ivy League gravy train—unlike the Quayles of the world who, despite their outstanding academic records, have had to settle for America's lowest-paying, lowest-prestige, lowest-powered jobs.

Of course, Bloomian authoritarianism may not have anything more than an analogical relation to Limbaughian authoritarianism: it would not be hard to imagine a future book entitled *The Western Canon the Way It Ought to Be,* but Bloom had nothing to do with the Republican sweep. Nevertheless, I do want to suggest that his success, like William Henry's, should be understood as a negative symptom of the general, if fitful, academic consensus that value is contingent and constructed: both Bloom and Henry spare no term of opprobrium in abusing academic literary critics, and this too is part of their appeal to the "common reader." Since this academic consensus is widely (mis)understood to entail the doctrine of cultural relativism, under which one book is as meritorious as any other, it would seem that the return of the Arbiter—he who knows what's good and why, and can place every form of cultural expression in its natural order—speaks to a widespread popular desire for clear-cut determinations of value. If I'm right about this, there's a real audience out there for the juridical form of literary criticism that unselfconsciously participates in the culture's general obsession with rankings.

I'm writing, however, to register something rather more critical than mere bemusement at Bloom's commercial success. Literary and cultural studies are indeed experiencing a crisis of evaluation, not just a "perceived" crisis; this is, in other words, a matter of intradisciplinary uncertainty as well as a question of public relations. And because we are uncertain, in ways Bloom is not, about what counts as good literary, critical, and theoretical work, it sometimes appears that the only criteria of evaluation we have are mercantile criteria: what's hot, what's selling, what's the newest latest. Or so says David Bromwich, in a remarkable passage in *Politics by Other Means* where, in rendering a detailed fictional account of a sham tenure meeting, he explains how vacuous are the criteria by which we promote professors in nonfields like "Postmodernism and Cultural Studies" (177). The tenure candidate is named "Jonathan Craigie," and he's a hotshot:

> The author of two books and editor of three collections of essays and interviews, the candidate in the last five years has delivered thirty papers at professional conferences—many of them deft and wide-ranging in the interdisciplinary way. His undergraduate courses are popular. They deal entirely with special subjects; for, armed with a counter-offer, Craigie negotiated early to sever his obligation to teach the department's bread-and-butter surveys. . . . The only recorded exception to the consensus occurred a year ago, when a pair of history students dropped [a graduate] seminar at midterm—because, they said, Craigie did not know the facts of the period implied by his title, "Postmodern Politics and American Culture, 1945–1990." (177)

Craigie's a poseur, that much is clear. But then things get worse: since Craigie's work is so post-everything (his next book "is a double biography of Madonna and 'a post-gendered novelist still to be named' " [178]), the tenure meeting turns entirely on the letters of his referees—as it must, writes Bromwich, because no one in the department can judge Craigie's work: "none had a language to describe or judge, let alone to praise or blame adequately, work that came in so curious a shape" (178).

And here's where Bromwich really goes to town, exposing not merely his fictional Craigie but every similar intellectual fraud perpetrated on the profession under the cover of professionalism:

> Of the six letters received, none came from a scholar who had done first-hand work on similar subjects; four, however, were written by eminences of theory, who could imply without quite asserting a meta-competence in several historical periods. (And, it was urged, what is postmodernism if not the theorist's topic par excellence?). . . . These carefully placed accolades were the fruit of assiduous networking: the cultivation, without ulterior intellectual aim, of influential cronies who are pledged to serve the advancement of one's career. Since Craigie had once devoted half a semester of a graduate seminar to the topic of networking, it was natural for this to form the first topic of an unusually frank and speculative tenure review. (178–79)

The crucial moment of the review comes when someone at last speaks up for traditional scholarly standards—the kind that prevailed in the pre-theory days, when tenure reviews were truly rigorous—and is stifled by one of the Eminences of Theory. The subject is Craigie's article "Alger Hiss Remediated: America's Left/Right, Vertical/Horizontal Control," and only one person in the room has the courage to say the essay isn't wearing any clothes:

> "This thing doesn't even have much information; I don't see any evidence that he reads the work he cites. In the Hiss article, he bases the whole thesis of 'image-consolidation' on a single discredited work of cultural history. He doesn't seem to have read any political history. He didn't look at the transcript of the congressional hearings or the trial. He didn't read *Witness* or *In the Court of Public Opinion*—nothing!"

A brief silence falls. Then—quietly, significantly—are heard the following grave words from the front of the room. The speaker is a recent high-profile recruit to the department:
"A couple of you people talk as if you were opposed to anything new." (180)

Craigie is eventually given tenure on the grounds that "if we turn him down lots of people will be asking why" (179) and "in a year or two, he'll have a great offer from Duke or Emory, and he'll leave" (180).

Part of what makes Bromwich's account so breathtaking is what I call the Gerontion Effect: the young neotraditionalist trying on the voice of the old man in a dry month. When first I read this account I couldn't help but think, in outrage, of the staid, solid scholars I've known who've run into trouble with incompetent tenure committees because their work was "too feminist" or had run afoul of one Gray Eminence or another (not a Theory Eminence, either). I thought of my three years as Illinois's placement director, watching one Ph.D. candidate after another fail to find full-time employment despite a terrific teaching record and a capable dissertation on a topic even Bromwich would recognize as legitimate. But once I managed to recover my breath, I realized I too had been taken in by Gerontion: for Bromwich's account had, in effect, asked me to compare Craigie's worthlessness (a worthlessness that can only be glimpsed from the privileged vantage point of disciplines with real standards, like history) with other contemporary scholars in English. And in so doing, Bromwich had briefly made me forget that in the golden pre-theoretical age he alludes to earlier, when lectures were not mere entertainment but learned addresses to "an audience of peers, who [were] relied on to know the material as well as the lecturer" (174), English professors basically got tenure if they could sign their names or prove they were carbon-based. Moreover, he'd also made me forget that those worthy people, tenured in the 1960s when times were flush, were now sitting in judgment of junior faculty and Ph.D. candidates who'd published and taught more just to get a job than most senior faculty had done to earn lifetime tenure thirty years ago.

Bromwich's shortsightedness is all the clearer if we contrast his book with a brief essay by George Levine that shares some of Bromwich's misgivings about the profession: "at the moment, the profession doesn't know or even want to know what it is," writes Levine. "[M]any of the best-known in our field are professionally interested in things that are only marginally related to English or literature," such as "cultural studies, film, gender, psychology, sex, pop culture, multiculturalism" (43). By contrast to Bromwich, Levine is, on the whole, sanguine about the field:

> English is a far more exciting and interesting, if qualitatively uneven, discipline than it seemed to be back then [when he entered it], and its current troubles have come about largely because it takes more risks, makes bigger mistakes, reaches out farther, welcomes more diversity, does better by women and minorities, and confronts its own assumptions more directly and perhaps even with greater sophistication. (43)

But as we've seen earlier, he cautions his colleagues that our health and our public support may depend on our commitment to a principle many in the profession do not in fact support: "We must examine the value of the literary and the aesthetic (even if only in the terms that Eagleton offers in *The Ideology of the Aesthetic*) if English, as a profession sustained by publicly and privately endowed institutions, is to survive" (43–44). What makes Levine's account both more plausible and more trustworthy than Bromwich's, I suggest, is that where Bromwich can only practice a form of ventriloquism, adopting the voice of the senior sage whose eyes have seen better days, Levine actually knows whereof he speaks: "When I got my degree from the University of Minnesota" in the 1950s, he writes, "almost all my colleagues, no matter how dumb they were, got at least three job offers" (43).

But then there was a third moment to my reading of Bromwich's tale of peer review gone wrong—the moment when I realized that Bromwich has a sound premise even though his account bespeaks his poor sense of disciplinary history (a nice paradox, since he exposes Craigie precisely for his fraudulent historical sense) and a truly odd sense of disenfranchisement (it is not every day that someone is appointed to head the Whitney

Humanities Center at the age of forty and responds with a screed about how such professional rewards usually go only to shallow networkers). Bromwich's caricature of tenure review may be outrageous, but it speaks to entirely legitimate concerns; and I myself would not want to bet the car on the proposition that a real-life Craigie could never win tenure. For we *don't* have a scale of values that can measure the work of people doing post-gender studies of Barbie dolls in the same terms as the work of people doing pre-gender studies of Hopkins's experiments in prosody. Nor can we plausibly compare *those* projects with articles in writing studies or books on recent Australian film—and yet we are routinely asked to do precisely this, for tenure reviews, fellowship applications, article submissions, and salary determinations.

I think it is true, then, that our professional evaluations do sometimes depend more on our sense of "currency" and "market value" than on our assessments of scholarly merit, and I think that in a bear market such as this one, when there are so few decent jobs for so many worthy candidates, the indefiniteness of our standards of merit warrants further discussion. The reason I won't echo the self-serving conclusion Bromwich draws from this, however, is that we need to remember (as Bromwich does not) that when our professional standards are uncertain, the appeal to "what's hot" can—and is—used to dismiss new work as often as it is used to praise it: *this has a certain flair,* one hears, *but it's insufferably trendy.* Or: *I suppose this is what's fashionable at Duke or Berkeley, but we needn't encourage it here.* Those who believe that peer reviews no longer contain such phrases are, I'm afraid, woefully mistaken. Indeed, because of the explosion of the field and the contraction of the market, the very idea of "peer review" has become increasingly problematic: on the one hand, the books of younger scholars are being judged by people who never wrote a book, and on the other, cultural studies work on Australian film is being judged by people who still don't like seeing the word "gaze" in film criticism (just as worthy "traditional" work is being judged by people who don't like *not* seeing the word "gaze" in film criticism). One way of phrasing this peerless paradox, as W. J. T. Mitchell pointed out nearly a decade ago, is to say that the discipline's official forms of pluralism are always bumping up against institutional limits on their operation. Asking "who is qualified to testify on the merits" of various scholars in the profession, Mitchell writes,

> With theory, feminist criticism, or Marxist criticism, the decision is much more difficult [than with eighteenth-century studies]. Chances are we have no senior person in the field, and that the very existence of the "field" as a respectable discipline is in doubt. If the existence of the field is granted, chances are that there is deep suspicion about the experts and standards to be consulted. The result is that everyone in a department feels qualified to judge persons in these fields, while in fact no one is really qualified. Chances are that no such appointment will be made, or that, once made, it will be more difficult to achieve tenure in the marginal, problematic field. (501)

Like Levine and Mitchell, I have no illusions that things were better in the days imagined by Bromwich, when everyone could be "relied on to know the material," because, for one thing, I have found from perusing past MLA programs that in all our professional gatherings from 1883 to 1963, only one paper was delivered on a black writer (Blyden Jackson gave a paper entitled "The Dilemma of the Negro Novelist" in 1953). More generally, unlike Bromwich, I would not want to purchase disciplinary consensus on our protocols of evaluation at the price of getting rid of all the writers, theorists, and intellectual movements that have done so much to problematize "value" in the past few decades. Until only a decade ago, for instance, our major professional journal would not accept essays on "minor" writers; the official policy of *PMLA* was that it would consider essays that took on "a broad subject or theme," "a major author or work," or "a minor author or work in such a way as to bring insight to a major author, work, genre, or critical method." And there is no advantage, it seems to me, in trying to return to some kind of disciplinary consensus that would bar "minor" writers from *PMLA* essays until they had been laboriously elevated to "major" status by critical work published elsewhere.

But what's impelled me to agree that our professional protocols of evaluation *are* in crisis, curiously enough, are the variously bitter and cynical accounts of the profession I've seen from people who have nothing much in common with Bromwich. Take, for example, this telling moment from a recent review essay on the Routledge *Cultural Studies* volume edited by Larry Grossberg, Cary Nelson, and Paula Treichler:

> The list of contributors on this book's cover is equally familiar, although precisely at odds with the kind of "collaborative" work it might herald in another economy—the economy this book keeps claiming, while it so successfully continues to operate within capitalism. Routledge's "Press Release" emphasizing the "prominent cultural theorists" writing in this book, and their winnowing of the list of names on the cover to headline the eight most prominent, makes clear the logic of advertising here: these are the celebrities, the *big* names who can draw an audience to a conference or readers to a book. (Langbauer 470)

Leaving aside the question of whether *all* the contributors to *Cultural Studies* were truly already "familiar" to everyone in 1992, let me suggest that this passage shares Bromwich's assumptions about the profession almost completely. These guys and gals in pomo and cultural studies aren't really scholars, they're more like celebrities; indeed, the essay is titled "The Celebrity Economy of Cultural Studies," and in it, Victorianist critic Laurie Langbauer penetrates to the empty core at the heart of the cultural studies enterprise, its dependence on (blindness to, reinforcement of, inscription within, etc.) precisely the forms of commodity capitalism it seeks to critique. "The marks of the celebrity economy in *Cultural Studies* may initially seem unremarkable," admits Langbauer, "because they are part of the customary economy of marketing under capitalism" (469–70). Ay, but there's the rub:

> All books do this. The problem is that the editors of this book treat "legitimate" cultural studies as if it did not. They indeed recognize the "significant investment opportunities" revolving around cultural studies in "academic institutions—presses, journals, hiring committees, conferences, university curricula" (1). Yet they ignore that it is within capitalism too that their own

conference is funded and their book published, and it is capitalism's laws that govern its marketing. (470) [1]

My guess is that one could not write such an essay unless one were
profoundly unsure of the intellectual merits of the book: for if we had
some measure of confidence that the work of these celebrities actually
merited their professional recognition (granting this even in the case of
the eight "winnowed" superstars, whoever they were), we wouldn't be
spending an entire review commenting more on the relation between a
book's marketing and its self-representation than on its contents. We
would think that the value of these critics' work merited their marketing,
instead of supposing that their marketing constituted their value. Once
again, that's not to say that we should want to return to the days when
everybody knew that a study of Christian imagery in the work of a major
author like Dryden was "intrinsically" more important than a study of
gender relations in the work of an unknown like Hurston (and therefore
no one wrote the latter). But it is to say that there's an extraordinary
level of cynicism in the industry these days, partly because our old
criteria of value have disappeared and we're not sure whether the new
ones have arrived yet, and partly because most new Ph.D.s' hopes of full-
time employment have disappeared as well, leaving us a cantankerous
profession with eight superstars drawing people to conferences and eight
thousand adjunct instructors teaching bread-and-butter surveys.

Langbauer's essay is not unique, nor is its seemingly gritty exposure of
professional "celebrity" applicable only to the nonfield of cultural studies.
On the contrary, the profession of whatever it is we do has recently given
birth to a new subgenre of such essays, wherein prominent critics such as
Donald Morton and Daniel O'Hara remorselessly critique the celebrity
economy by which other critics have become "superstar 'free agents'"
(O'Hara 43), construing lesser-known faculty and students, in Morton's
more-cynical-than-thou locution, as "celebrities-in-waiting" (144).
What's driving these essays, however, is not merely the intradisciplinary
explosion that's made it all but impossible for us to judge the relative
worth of a new close reading of Robert Frost, a new theoretical wrinkle
in writing studies, and a new critique of the suturing of the subject in

the political imaginary of the post-(techno)logical sublime. Rather, as O'Hara's, Langbauer's, and Morton's essays variously make clear, it's our intradisciplinary explosion *in tandem with* the collapse of the job market. What's happening to our professional protocols of value is that they're being squeezed by a system whose ideal image of itself promotes theoretically sophisticated, interdisciplinary work in extraliterary studies but whose material base is shrinking as fast as its superstructure is expanding. To put this impasse in historical perspective (since it no doubt will be counterargued that the job market has been awful before), it was roughly twenty years ago that Richard Ohmann pointed out, in *English in America,* that the profession of English studies thought of itself as doing criticism and theory but was in actuality devoting half its courses to introductory composition. In the 1990s, by contrast, the profession devotes *more* than half its courses to composition, faces a rising number of Ph.D.s together with a declining number of tenure-track (or even full-time nontenurable) positions, and thinks of itself as doing everything from recent French philosophy to analyses of popular music to new editions of the C-text of *Piers Plowman.*[2]

I would have no difficulty with this profusion of material for study—nor, I wager, would many of my un-Bromwichian colleagues—if it were not coincident with the collapse of the job market. It's not that I believe the job market would improve if we would all agree to stick to reading *Piers Plowman* and leave the hip-hop to writers for glossy magazines; the market works by variables that have nothing to do with the profession's intellectual interests. On the contrary, the profession's intellectual interests can often be the dependent variable (dependent, that is, on the market)—and that's the problem, particularly for our graduate students: the discipline thinks it's going from literature to culture, and the market tells us we're going from literature to technical writing.

One effect of this conundrum is that students have little idea how to value (or how to gauge whether the profession will value) a dissertation on slasher films vis-à-vis a dissertation on sprung rhythm vis-à-vis a dissertation on penal codes in nineteenth-century Australia vis-à-vis the ability to teach four sections of comp, the History of the English Language, and maybe a Shakespeare course every other year. They suspect, correctly, that a truly singular dissertation on slasher films *and* Australian penal codes will win them more "attention" from search committees and publishers than a careful reading of Hopkins's prosody—but then again, they also suspect that because there are only two jobs in the country for writers of such dissertations, they're going to wind up doing piecework comp courses for the local community college at $900 per while they try to send the book off to Routledge. But even should they publish that book with Routledge or anywhere else, they still have little reason, often enough, to expect full-time employment; and when a good teaching record and a published book cannot get you hired at a university—in a system where indifferent teaching and a published book was once sufficient for tenure and sometimes even for promotion to full professor—it is no surprise to find widespread professional confusion and cynicism about what the profession professes to value. The recent experience of one of my students is perhaps a leading indicator of what the coming years will look like: having published two essays in important journals while temping at a variety of small colleges, the student finally got a book manuscript accepted by a prestigious university press, only to be told—*after* completing nine months of revisions—that the marketing department had nixed the book on the grounds that it wouldn't sell widely enough. Had the book been accepted, the student might have had a chance at landing a new kind of job offered by a nearby college: a 4/4 (four courses per semester) instructorship leading (for the worthy candidate) to tenure *without* promotion. In other words, a lifetime instructorship without hope of advancement, regardless of the quality of one's teaching or scholarship. Without a book in hand, however, the student had a poor chance of competing against the rest of the applicant pool for this rotten job, since so many of the other applicants, according to the search committee, had already published one book or more.

The message to current graduate students would therefore appear to be merely overwhelming: you are responsible for mastering many times as much "scholarship" as your seniors were; your "field" runs from Chaucer-Shakespeare-Milton to Andrew Ross on biotechnology and Teresa de Lauretis on spectatorship; and you will be expected to contribute substantially to the "field" well before you've been credentialed to join it.

For new Ph.D.s the message is slightly different: if you're unemployed, misemployed (working the night shift at the copy shop), underemployed (teaching 4/4 loads of rhetoric and composition), you need to publish all the more, even though your working conditions do not permit you to do so. And then those of you who are lucky enough to scramble into a tenure-track job can *then* begin the "probationary period" during which you will be assessed for your ability to contribute to the reams of scholarship you're also (still) responsible for mastering.

To put this starkly, we are at the point at which many entry-level jobs require a published book; indeed, for a few Illinois Ph.D.s who've been job searching for the past few years and temping at various colleges in the meantime, a published book now seems the only thing that will make them "attractive" enough for *any* full-time position. This is not hyperbole. Our graduates who received their Ph.D.s in 1990 or 1991 and who've published substantial articles in major journals are attracting very little attention from potential employers, even when their research is considered good enough for reputable journals and prestigious university presses. My hunch is that these Illinois graduates are not market anomalies, and that many of our current students who'll hit the market in the late 1990s will likewise have to produce a book before they'll be considered for the tenure track.

I see two ways in which the book-for-job criterion has heightened tensions and worsened working conditions in the profession. First, it has further alienated secure senior faculty from their jittery juniors and vice versa: at some institutions, senior faculty who won tenure twenty-five years ago, in vastly different circumstances, are now teaching alongside assistant professors and graduate instructors who've published more stuff in three years than they wrote in thirty. And even when it does not issue in intergenerational tension, the precipitous rise in the profession's publishing requirements can make peer review difficult—as is always the case when "peers" are operating on very different professional standards.

I don't mean to suggest that faculty who publish widely are necessarily more valuable or more "productive" than relatively unpublished faculty who may or may not be terrific teachers; on the contrary, I want to suggest that any profession that encourages its own publishing glut paradoxically devalues publishing precisely by overvaluing it. And since I've published a fair number of things myself in recent years (and can thus be taken as a symptom of the very problem I'm describing), let me back up and remind my colleagues how drastically the standards have changed in only ten years: in the fall of 1988, when I began looking for a job, I had published only one essay—and had not yet presented a paper at a professional conference. Almost none of my fellow Ph.D. candidates at Virginia had accomplished anything more; the vast majority had published nothing, and very few of us had given papers. The twenty-five Virginia candidates who went on the market that year garnered a total of 175 interviews. A mere two years later, when the bottom fell out of the market dramatically (and, from the looks of things, permanently), a total of forty Virginia candidates picked up only thirty-six interviews—and from that point on, graduate students knew, as I had not known in 1988, that it would be folly to test the market without having compiled a substantial record of professional achievements.

If professors and students alike are thus encouraged to value quantity of production over quality, then Ancients tenured in 1969 can find themselves stranded in a profession whose rule changes have taken the rug out from under them; as a result, all too often, they wind up considering the whole business of professional criticism a sham. For their part, Moderns awarded their Ph.D.s in 1993 can find themselves asked to write more stuff just for an interview than some of their interviewers wrote for tenure, and as a result, they all too often wind up considering Ancients as nothing more than blocking figures. Each side then construes the other as the source of all the profession's ills, and it certainly doesn't help matters, in the current political climate, that most of the tenured Ancients are white men, and many of the underemployed Moderns are, generally speaking, Others. Thanks to that demographic spread, young white men can blame the crunch on the Others, and the Others can blame it on old white men, and the middle-aged white men can ask, where are all the great critics of yesteryear? Has anybody here seen my old friend Northrop?

My point is that there's an increasingly tenuous relation between our discipline's expectations for publishing and the jobs our discipline characteristically offers its new Ph.D.s. Indeed, it's increasingly common that Ph.D.s will be asked to produce a publishable book-length manuscript, a couple of articles, and a smattering of conference papers in order to be considered for jobs in which research will play no significant part, or will actually be made nearly impossible by crushing teaching loads. Consequently, I have had to counsel candidate after candidate who was amazed to learn that s/he would only be considered for an interview, even by the smallest colleges, on the basis of their research and publications—only to find themselves looking at a 4/4 teaching load in a department that could not support junior faculty research. I'm not saying this practice is ubiquitous; some colleges whose search committees expressly valued teaching over research proceeded by contacting our candidates for interviews without having requested a writing sample. But if a department with a 4/4 load in basic instruction should require further research for tenure, then the new Ph.D. is faced with a demand s/he is not likely to meet; and if the department actively discourages research, and never wants to see evidence of the dissertation again, then the new Ph.D. has been seriously mis- or underemployed, with little chance to move elsewhere in a profession where "vertical" advancement depends entirely on publication. To date, my advice to Illinois Ph.D.s in this situation has been this: if you're seriously mismatched with your school, take every spare moment you can and build the machine in the basement (a book manuscript, a series of articles) that you hope will enable you to escape. The problem with this advice, of course, is that every misemployed Ph.D. is following it, thereby driving our publication requirements higher still.

Before we congratulate ourselves for moving from literature to culture, therefore, we should at the very least ask how this new *Selbstverständnis* meshes or conflicts with the brutal realities of the job market; it may be plausible, for instance, to think (as people like James Berlin—or, in a different vein, Henry Giroux—have done) of cultural studies in English more in terms of its relation to rhetoric, pedagogy, and composition classes than as a competitor with "literary studies" for already scarce disciplinary resources. Conversely, it may be possible to envision an American cultural studies as an inter- and antidisciplinary field of endeavor by practicing it in nondepartmental units, such as humanities institutes or African American studies programs; this option would allow students and young faculty in cultural studies to attempt *real* interdisciplinarity rather than the interdisciplinary waterskiing that is all most ordinary graduate students have time for, but the likelihood of realizing this possibility will depend simply on whether universities will want to hire students so trained. In any case, the "success" of our disciplinary shift from literature to culture is not something that will be decided by journal articles or books—this one included. Rather, it will be determined largely by the availability of jobs, which in concert with the dissemination of cultural studies will in turn determine the extent to which English departments will remain places that study and value "literature" as most of the literate culture understands it.

Two conclusions follow from this line of argument: one has to do with English's formal protocols of value as they are perceived outside English, and the other has to do with the future of literary study. For as long as I've been involved in the profession—beginning with my entrance to graduate school in 1983—I have thought of, and experienced, English's heterogeneity as something admirable, exciting, and valuable. Indeed, it's for that reason that I never understood "disciplinarity" as the dirty word it is sometimes taken for (as in Ellen Rooney's article on cultural studies, "Discipline and Vanish"), since I had the good fortune to be working in a discipline notable for its vigorous interdisciplinary importation and exportation of methodologies. But the flexibility of English is as often ridiculed as admired by scholars in other disciplines, and the more an outside observer values disciplinary stability, consensus, and "performativity" (in Lyotard's sense), the more of a liability English will seem to him or her. Indeed, to judge by a recent book that defends the American university partly by distancing its authors from the "dangerous" loonies in English, some of our fellow humanists construe our disciplinary messiness as a sign that English should (or does) not exist. In *Up the University,* Robert and Jon Solomon write,

"English" is not a natural subject, but a dangerous amalgam of functional service courses (teaching two of the basic 'three Rs'), an arbitrarily truncated literature program (which seems to include only writing that does not require knowledge of a foreign language), and an often arrogant collection of literary theorists, who are typically at war with each other as well as with their more down-to-earth colleagues. They make their reputation by attacking and alienating everyone else in the department. There is a very real question why and whether the university should continue to have an English department at all, but for now, we want to make a much more modest claim—that the university is not the English department, and the embarrassments and anecdotes of a few notorious departments should not be taken as the nature of the university itself. (296)

It may seem odd that we be faulted for our disciplinary disorganization (not to mention our constitutional arrogance) at the same time we are faulted for not teaching literature in languages other than English, but perhaps this passage is motivated more by fiscal than by intellectual concerns: whatever else they are doing, the Solomons are basically proposing that English take the first and deepest cuts when the college budget is on the block. It is important for this reason that these criticisms come not from Kimball, Sykes, Cheney, and Company but from colleagues across the quad, housed in departments of philosophy and classics.

My point, obviously, is that should arguments like this carry the day with deans, donors, trustees, and local legislatures, the scarce resources for work in English could get still scarcer. Again, I'm not suggesting we stick to *Piers Plowman* in the hope that the dean and the trustees will look kindly on English and return to us all the faculty lines they've "absorbed" in the last few years. The point is to *contest* these accounts of the field, and at the very least to come up with convincing rationales— that is, rationales that will not only convince *us* of the value of what we do with literature (though we do need these), but will also be intelligible outside the discipline—as to why it is that English values its disciplinary heterogeneity, and how it is that the value we place on heterogeneity does not so preclude us from making determinations of scholarly value that we are forced, like Bromwich's fictional tenure committee, to rely simply on scholarly commodity fetishism as a substitute for "value."

I said that my second point had to do with the future of literary study. The turn from literature to culture unfortunately comes at a time when American universities are seeking to divest themselves of "unproductive" fields of study and modes of inquiry. As English is increasingly compelled to "instrumentalize," that is, to devote ever more of its course offerings to basic writing instruction, literary study will indeed be found vestigial if we do not contrive convincing justifications for it. In recent years it has sometimes seemed to us sufficient to point out that most of the claims made for literature can be made for "extraliterary" works as well, just as it has seemed satisfactory to note that the ancient study of rhetoric was much more capacious than the New Critical study of "literariness," and that "literature" has come to mean "belles lettres" only in the past two or three centuries. These arguments are right, I think, and they have done salutary work in unravelling the brief but powerful (and powerfully deceptive) disciplinary consensus of the 1940s and 1950s that literature was an object for study unlike any other object of study. Yet even the most narrowly "literary" works really can defamiliarize the familiar and renew perception; compel readers to imaginative sympathy, disgust, ecstasy, terror; train young adults to attend to the subtleties of language, the rhythmic variations of verse, and the power of rhetorical hermeneutics; lengthen children's attention spans to the point at which they can understand how Frank Churchill's letter finally exposes Emma Woodhouse's inability to read *Emma*'s multiple subtleties of language; and, in rare cases, make undergraduates curious enough to keep reading after they graduate. And when critics on the cultural Left point out that none of this is necessarily inconsistent with the project of giving students mere ideological obfuscation or training them in quietism, my impulse is to agree—and then to suggest in return that if one desires guarantees that one's teaching and writing can never be put in the service of ideological obfuscation or quietism, one would be better off not wasting time with the humanities in the first place.

Besides, if these formalist rationales aren't enough to justify the discipline, you can always come across one of those literary texts that so trouble the Right, whose author may in fact have been more savvy about gender, power, empire, hegemony, or sexuality than your average associate professor in the Midwest. For that matter, those critics who are engaged in "archival" work with the neglected writers of the past or "advocacy" work with the underread writers of the present have an important stake not merely in maintaining but in expanding literary study *as* literary study: neither such critics nor their client-authors will be well served by a disciplinary regime that jettisons "literature" as a spent vehicle no longer able to assist in cultural study's further expansion.

For all his commercial success in 1994, Harold Bloom managed only to offer the claim that aesthetic contemplation of the beautiful is an end in itself; perhaps we theory-addled multiculturalists, we whom Bloom calls "displaced social workers," can do better. At the very least, we should not allow Bloom's solipsistic defense of the field to stand as the only one in public view. Whenever we are castigated as politicos and philistines intruding on the purity of the aesthetic experience, we need to remind our accusers that—as one theorist once put it—literature not only delights but instructs, at least every once in a while. It is a shame that arguments such as these are now more prevalent among cultural reactionaries, embittered aesthetes, and right-wing flacks than among the leading figures in departments of English; reactionaries and flacks have done much to make these arguments suspect, and Bloom himself has done more than anyone to discredit the idea that value judgments in the field of English can be something other than expressions of idiosyncratic predilections.

As for my own predilections, I can say that whenever I read a professional reactionary like Robert Brustein maligning the "victim art" of choreographer Bill T. Jones on the grounds "that reality without artistic transcendence turns us all into voyeurs, that the visionary gleam rather than any narcissistic glitter is what redeems American art" (10), I want to run screaming from any discussion of "aesthetics" or "transcendence." But we should not let the hypocrisy of flacks or the arrogance of Blooms deter us from making the case for the value of literary study *and* the

forms of literary scholarship that have so broadened—and troubled—
the field. It is time we took up the challenge to offer evaluative criteria
that will answer, rather than merely denounce, those of our detractors;
and we need not devise Top Ten lists or declare the inviolability of the
aesthetic in order to make the case that the power and pleasure of
imaginative literature should be among the things our discipline values
most highly.

NOTES

1. I am, of course, indirectly implicated by Langbauer's review essay,
   since I also reviewed *Cultural Studies* (for the *Village Voice Literary
   Supplement* in April 1992), thus helping to commodify the book still
   further. Accordingly, I think I should explain the circumstances of that
   review here. I was originally given two thousand words, or about a
   page and a half of *VLS* space, for my review; but when I turned in that
   draft in January 1992, my editors became more interested in the
   subject matter of the book, and suggested that I expand the review,
   narrate a brief history of cultural studies, comment on the conference
   proceedings (since, as they knew, I had attended nearly every session
   of the conference), and speculate on where cultural studies might go
   from here. In short, the *VLS* editors, often more enthusiastic than I
   was about some aspects of the conference and the book, decided to
   devote the entire April issue to cultural studies, and to use my now
   ten-thousand-word essay as a kind of flagship piece. Though I did
   comment, in passing, on the "academic celebrity" phenomenon (one
   could hardly comment on the conference proceedings without doing
   so), I will leave it to others to decide whether I did so fairly, or did so
   to the exclusion of discussing substantive matters about the conference,
   the book, or the field.

   In March 1992, the *VLS* ran a preview of its April issue, closing it
   with a tag line on *Cultural Studies:* "if you plan to continue living in
   America, read this book." I have since seen this line in Routledge's
   promotions for the book, run together with a passage from the closing
   paragraph of my essay, in which I wrote that cultural studies would be
   contested terrain in the 1990s and that *Cultural Studies* will be "the
   most capacious text in the fray." I have since been accused of making
   the reading of *Cultural Studies* into a residency requirement for U.S.
   citizens, and I want to make it known that I never wrote any such
   thing. I merely said the book was really big, and that no serious or

curious reader should ignore it. Looking back on the review some years later, I find neither proposition especially questionable.

2. The extent to which Levine's argument is an update of Ohmann's can be gauged by the following: "The two functions of English departments that institutions and the culture as a whole endorse, and pay us for, are perhaps the two to which we as research faculty members are least committed. One is the teaching of writing as a basic skill that all educated people need to acquire, and the other is the teaching of literature as it is widely understood by those who don't make the study of it their profession" (44). Likewise, at the end of the essay, Levine comes to the sound, Ohmannian conclusion that "we must learn to build departments whose interests and objectives are less at odds with their immediate public responsibilities" (45). For Ohmann's analysis of the relations among the teaching of basic writing, the prestige system of the profession, and the public rhetorics of justification for literary studies, see *English in America*.

5

STRAIGHT OUTTA NORMAL

NONPROFIT FICTION PUBLISHING ON THE MARGINS

I argued in chapter 1 that contemporary literature in English has been strangely and unconscionably overlooked in our current debates over literary and cultural studies. Since the previous three chapters have not touched on literature at all, except to gesture, at the close of the last chapter, toward the importance of literature in a cultural studies curriculum, readers may well be asking by this point why I should have made such an argument in chapter 1, and how, if at all, it informs my own practices as a reader, writer, and teacher. I am no expert, after all, in Michael Ondaatje or Buchi Emecheta; indeed, if English departments became departments of "literatures in English" tomorrow I would be among those faculty who would need to begin reading more widely in contemporary literature in order to stay current with the discipline.

What's more, I know full well that whatever reputation I have as a critic has not, for the most part, depended on my skills as a close reader of literary texts. This state of affairs has begun to strike me as odd, not only because I spent my undergraduate and graduate years doing nothing but producing close readings of literary texts (I still have a sinuous argument about Book IV of the *Faerie Queene* if anyone's interested), but particularly because my courses at the University of Illinois have been, with only two exceptions, courses in American and African American literature. At the end of my first five years at Illinois I calculated that I had assigned over sixty different American writers, from Nathaniel Hawthorne and Harriet Jacobs to Willa Cather and Henry Roth to Toni Morrison and Richard Powers. To my surprise and delight, I had even taught a couple of courses in contemporary literature, usually under the heading of Postmodernism and American Fiction. I had even published a first book, *Marginal Forces/Cultural Centers,* in which I described, among other things, the gap between "creative writers" and "critics" in the academy. But in my own life, the gap between my writing career and my teaching of literature was only growing wider.

Then one day in 1992 Charles Harris of Illinois State University sent me a long letter in which he took up the argument of *Marginal Forces* and gave me a chance to do something about it: inviting me to contribute to a special issue of *Critique* that would be devoted to the work of nonprofit presses in the United States, he asked whether I would consider the work of Fiction Collective 2 as a place where contemporary "creative writers" and contemporary critics/theorists might meet. I accepted the invitation, and spent many months in 1993–94 reading here and there in the Fiction Collective 2 list, getting a feel for the terrain and trying to figure out what I could say about the cultural status of nonprofit "avant-garde" publishing houses that were supported in part by federal grants, private enterprise, and university subsidies. Was it possible, I wondered, to have an academically subsidized avant-garde, or was the concept of the "avant-garde" itself a modernist hangover from which we would do well to recover?

In March 1997, these questions were answered dramatically when the cultural status of Fiction Collective 2 was placed firmly on the national agenda—and not by me, either. While cultural studies and most of the literary academy were ignoring FC2, Representative Peter Hoekstra (R.-Michigan) was discovering that the press sometimes publishes sexually explicit material, some of it involving intimate physical relations between woman and woman. The guilty text in question was a collection titled *Chick-Lit 2: No Chick Vics,* edited by Cris Mazza et al. (about whom I'll have more to say below), and the reason it was important was that it gave congressional Republicans the ammunition they wanted to defund the National Endowment for the Arts once and for all. The argument, of course, was that taxpayer dollars should not be spent to help produce sexually explicit representations of same-sex couples. There is an obvious (and entirely democratic) counterargument to this position, a counterargument that has nothing to do with likening Hoekstra and his allies to Nazis who crusade against "degenerate" art. It goes like this: despite what Republicans want to believe, same-sex couples who enjoy sexually explicit representations in literature (not to speak of heterosexual couples who enjoy sexually explicit representations of same-sex couples) are actually part of the current population of the United States. Indeed, they make up far more of the citizenry of the Republic, in proportional terms, than the portion of the national budget that goes to Fiction Collective 2: since FC2 received $25,000 in fiscal year 1996, or approximately .000000017 of a total budget of $1.5 trillion, and .000000017 of the current U.S. population (250,000,000) happens to be *four,* it's probably safe to say that FC2's portion of the national budget does not exceed its popular mandate.

All of a sudden, the 105th Congress had found its Mapplethorpe—except that *this* time, in an unprecedented move, elected officials were targeting an entire publishing house rather than a specific text or artist. Even the fact that *literature* was under attack had a kind of exotic or antiquarian feel to it, as if we'd been catapulted into fundamentalist Iran or whisked back to the days of the U.S. *Ulysses* trial in 1933: our national debates over sexually explicit representations have been almost entirely taken up with *visual* pornography since the 1960s, just as most pornography tends to be visual as well (so much so that it is almost unheard of to come across an "adult books" store that actually does a brisk business in quality fiction). At this writing I am not sure that FC2 or the National Endowment for the Arts will survive this latest attack, so I do not want to sound glib or facetious in describing what I want to call the long-awaited Return of Comstock. Likewise, even though there's much to hope for in Representative Hoekstra's attack (perhaps Republicans will begin reading more widely, and perhaps FC2 will benefit from the general publicity), still, there's a real danger here that the always vulnerable National Endowments will be eviscerated by the forces of privatization and plutocracy—and, more to the immediate point, there's every chance that the Fiction Collective 2 list, in all its diversity and richness, will get lost in the Capitol Hill Shuffle. That list, as I hope to show in this chapter, is in some sense "subversive," and it contains plenty of sexual explicitness, too—more than enough for any Helmsman or Orrin Hatchetman (or their henchmen) in the U.S. House or Senate. But as usual, the material under scrutiny here is more complicated and rewarding than most of our democratically elected cultural fascists can account for.

The Fiction Collective was founded in 1974 at Brooklyn College by Jonathan Baumbach, Steve Katz, Peter Spielberg, and Mark Minsky. The year before, Richard Kostelanetz had published his incendiary book, *The End of Intelligent Writing,* which basically accused the entire publishing industry of being a closed shop of logrolling think-alikes—*and,* more accurately, pointed out the extraordinary coincidence between the amount of ad space bought by various publishers in the *New York Times Book Review* and the amount of review space allotted to those publishers by the editors of the *Review.* The Fiction Collective, from its inception, meant to shake up that arrangement. Through the 1970s the collective thrived, turning out wave upon wave of innovative, exciting fiction, garnering critical acclaim not only from variously *avant* folks like Robert Coover and Jerome Klinkowitz but also, on occasion, from *Newsweek* and the *New York Times* to boot. Raymond Federman's *Take It or Leave It,* Ron Sukenick's *Long Talking Bad Conditions Blues,* Baumbach's *Chez*

Charlotte and Emily, Spielberg's *Crash-Landing,* Clarence Major's *Reflex and Bone Structure,* Fanny Howe's *In the Middle of Nowhere*—all were Fiction Collective books, and all helped establish FC's reputation in the publishing world, in the academy, and among "serious" writers and readers.[1] Fiction Collective writers found champions like Rachel Blau du Plessis, Toni Morrison, and Morris Dickstein, and the original FC group itself was unified enough to appear as a "school," with an identifiable brand name and trademark: whether it was called "surfiction" (Federman) or "post-contemporary" (Klinkowitz), FC fiction was distinguished by its bad attitude toward *réportage,* its narrative reflexivity, its typographical oddities, its penchant for addressing readers directly—you know, its general preference for ludicity over lucidity.

But by 1987, FC was down to publishing only one or two books a year from the five to six a year of the 1970s. Part of the problem was the Fiction Collective itself: originally a small, close-knit boys' club dedicated to publishing weird writers frozen out by the onset of MergerMania and MallFiction, the collective had grown downright unwieldy over a decade and more—partly because each new writer became part of the collective upon publication of his or her work, an arrangement that eventually presented the group with the headache of coordinating something like forty voting members. The collective also had its image problems, as well, in the form of an annoying habit of representing itself as more-*avant*-than-thou, the vanguard's vanguard—and aside from people like Clarence Major or Fanny Howe, its constituent writers were as overwhelmingly white and male as any corporate boardroom. But at the same time, the collective would sometimes be the publishing world's answer to Broadway Danny Rose, taking on talent that no one else would touch—until the talent got hot and moved on to the big time.

Fiction Collective 2 was therefore born from FC's ashes in 1988, and is now run by the leaders of the new school, Ron Sukenick and Curtis White. Like many other small presses, FC2 now publishes a very healthy ten to twelve books a year, including the new Black Ice series (inaugurated in 1993) and a series begun in 1991, On the Edge: New Women's Fiction. Each year FC2 also publishes the winner of its national fiction competition, as well as the annual Charles H. and N. Mildred Nilon

Excellence in Minority Fiction Award. The current collective writers no longer constitute an identifiable school, but at the same time (and for the same reasons), what FC2 has lost in brand-name clarity it has gained in heterogeneity and vitality.

Until it became material for congressional testimony, however, the FC2 lineup was not nearly so newsworthy as was the original FC cast. Even though FC2's brand of experimental fiction is, by and large, every bit as good as experimental fiction ever was (say, since *Tristram Shandy*), and the rejuvenation of FC2 is an extraordinarily bright sign for dissidents, punks, and strange writers everywhere, it's so far been much less visible to the literati than was its earlier incarnation—largely thanks to the fact that the mainstream trade press reviewing apparatus has shriveled drastically in the past fifteen years. If Richard Kostelanetz thought it was hard for an independent publisher to draw major review attention in 1973, he surely knows that by now it's nearly impossible. Nonprofit publishers get even less space from generalist reviews than does contemporary literary theory, which means you probably won't find *Newsweek* or the *New Republic* trumpeting any FC2 titles in the near future. But like contemporary theory, FC2 work is finding more responsive (and responsible) audiences in the pages of the alternative press—from the *Village Voice* and the *San Diego Reader* to the more "specialized" readerships of *Science Fiction Eye* and *Maximum RocknRoll* (the nation's largest and longest-lived review of punk and post-punk culture).[2] And the market for FC2 books, though considerably less "academic" than the market for theory, is roughly the same size: new books usually appear in runs of two or three thousand, which means that FC2 work is very likely competing for some of the same .00001 of the American public that buys the work of those jargon-addled theoreticians and talented small press poets and fiction writers no one's ever heard of (.00001 is only 2,500 people, but don't forget that .00001 of the national budget is still a hefty $15 million). If there tends to be a good deal of hostility, suspicion, and mutual indifference circulating among the *avant* FC2 crowd, the theory-laden MLA crowd, and the trad-fiction MFA crowd (and the volume of anti-academic rhetoric among trad and nontrad writers is but one index of this), that's not only because their intellectual

117

enterprises are often at odds; it's also because they're all fighting for the same small subculture of active, intellectually engaged readers—whether in the pages of the *TBR,* the *AWP Chronicle,* or *Maximum RocknRoll.*

As Sukenick has it, the transition from FC to FC2 is not so much a change in style or cultural location as a change in the collective's approach to the idea of a "popular" audience: whereas FC, like New Directions, tried to exist against or despite the mass market, FC2 is actually trying to break *into* the mass market. Sukenick's analogy—and you can find this in the promotional material for Black Ice—is "the recent explosion in the alternative music scene."[3] But that analogy only begs the question: sure, it would be wonderful if FC2 emerged from the garage, like Nirvana, to reach the mass market unmodified and reshape the industry. But the "explosion" in "alternative" music is also an explosion in the meaning of the word "alternative"—as evidenced by the fact that the turgid Pearl Jam and the blandomatic Bud Dry are marketed as "alternative" bands and beers. So perhaps the challenge for marginal cultural producers in the 1990s and beyond will be to come up with an alternative to "alternative." All the same, FC2's relation to alt-pop and alt-mass does indeed signal an important cultural shift in independent publishing. Small quality presses like New Directions were defined by the romantic-modernist dream of preserving Great Books from the detritus of mass culture and the corruptions of the marketplace; the original Fiction Collective, too, participated in that dream, trying to create a small enclave in which "the market" would have no force over the production of new literature. FC2, by contrast, characteristically tries to exploit the logic of the market (with its new series, its streamlined structure, and its wider distribution), and bore into the mass from within. This strategy suits its younger Avant-Pop writers just fine, since it enacts the strategies pursued by their own writing. What's odd about this configuration, though, is that it leaves us with an "avant-garde" that doesn't actually take up the "pure" cultural position of an intellectual vanguard. If you're sympathetic to Andreas Huyssen's cogent account of postmodernism as a "post-avantgarde" culture, then, the distinctive thing about FC2 is that its texts and readers are as likely to show up on Internet or post-punk fanzines as in small cafés, disseminated as circuitously and as unpredictably through the culture as is postmodernism itself.

In 1980, Peter Quartermain's *Chicago Review* article on the Fiction Collective noted that most of its leading writers were academics or authors of scholarly studies, and that "one would expect it to take a theoretical or programmatic approach to the art of fiction" (67). Today, although FC2's work can still be called "theoretical" (but not necessarily "programmatic"), its authors tend to describe themselves as artists, law students, dog trainers, playwrights, musicians, drifters, poets, and, on occasion, teachers and academics too. Some of them, like Cris Mazza and Richard Grossman, publish with both FC2 and other small presses (in Mazza's case, Coffee House; in Grossman's, Graywolf). Others, like Mark Leyner and Gerald Vizenor, published with the collective in the 1980s and have since "crossed over" to wider audiences.

Leyner's first book, *I Smell Esther Williams* (1983), offers twenty-six short somethings; I'd call them "stories" but for their cumulatively antinarrative effect. The opening piece, appropriately titled "Launch," gives some idea of the whole: beginning with "I've given the raft with the woman you've been waiting for a little push so you should be receiving her any day now" (3), the four-page piece quickly complicates the idea of woman-as-exchange. "If it is late at night," Leyner writes, "we pretend that we have lost the right to vote and that we have been sterilized by missionaries" (4). Leyner has the courage to insist "it's about time that the young American poets took Marianne Faithful off her pedestal" (78), to write solemnly of "the anonymous citizens of Targetgrad" (185), and to proclaim that "the slightest pang feels isometric and giddy and wanton like so many handfuls of hair, because I have drawn asbestos dust into my lungs and drunk the milk of michigan and dragged you out of an impending marriage for twelve hours in plain night" (35). But it seems somehow disappointing, not to say smoothly orange-rind, that the title story itself—at twenty-seven pages by far the longest thing in the volume—neglects even to mention Esther Williams. But then, it *was* Leyner's first book.

If Leyner offers us what Fredric Jameson called (in another context) "surrealism without the unconscious," Gerald Vizenor's second novel and only FC2 title, *Griever: An American Monkey King in China* (1987), gives us surrealism amid myriad double consciousnesses: the Native American protagonist, Griever de Hocus, is "a mixedblood tribal trickster, a close relative to the old mind monkeys" (34) who loves "women, heart gossip, stones, trees" and collects "lost shoes and broken wheels" (21). He's on something like a Fulbright, teaching in the China of 1983; his opaque and sinuous narrative begins in hallucinatory visions of fire bears, black opal rings, and birchbark manuscripts, and ends with the brutal death of Hester Hua Dan, whose father, Egas Zhang (Griever's supervisor, as it turns out), murders her when he learns she is carrying Griever's child. What begins in dream and comic rebellion—as when Griever liberates chickens from the market square in Tianjin and replaces Chinese patriotic music with John Philip Sousa on the campus loudspeakers—gradually turns deadly serious, as Griever interrupts an execution caravan (freeing a drug dealer, a thief, a prostitute, an art historian, and three rapists), disrupts the patriarchy that kills Hester Hua Dan, and finally escapes with Kangmei (Hester's half-sister by an American father) on an ultralight aircraft shipped to him from the reservation of his birth. The birchbark manuscript, filled with occult writings that remind Griever of "the tribal medicine scrolls from the reservation" (18), turns out to contain nothing but recipes for blue chicken. The personal becomes the political, the cryptic becomes the quotidian, and the trickster runs afowl of the state. Of Vizenor's trickster tales, *Griever* is often considered the most difficult; but among FC2 offerings, it looks right at home.

FC2's 1990s list, as best I can make it out after perusing roughly two dozen titles, more or less falls under two major headings, neither of which is definitive; any individual author will exceed them in some way. On one side there's the forbidding narrative experimentalists, carrying on one or another modernist tradition regardless of whether the AWP, Time Warner, or Vintage Contemporaries acknowledges its existence. On the other side you have the more diffuse Avant-Pop bunch, less inclined to fine writing than to visceral affect. (The Black Ice series tends to feature the Avant-Pops; Black Ice books are small and cheap—backpocket size, $6 or $7, just right for the coffeeshop habitué on the go, tuft of chin hair optional.) Under the neo-modernist heading I put writers like Rosaire Appel, Kenneth Bernard, Lou Robinson, Jacques Servin, D. N. Stuefloten, Yuriy Tarnawsky, and Kathryn Thompson; under Avant-Pop I'll put Mark Amerika, Ricardo Cortez Cruz, Eurudice, Bayard Johnson, Philip Lewis, and Derek Pell. And then for writers like Omar Castañeda, Cris Mazza, and Richard Grossman, you can make up your own designation.

I'll confess to being more comfortable with the neo-mods than with the Avant-Pops, but I also have to admit that the A-Ps usually provide the steeper and faster roller-coaster rides of the two. Bayard Johnson's *Damned Right* (1994) seems to me in many ways the most promising of the crop so far. Mixing Kerouac's manic energy with Nathanael West's apocalyptic black humor, Johnson has written an all-roads-lead-to-L.A. novel that manages to be exhilarating and poignant even when it's recycling a few on-the-road American-lit clichés. The narrator's ambition is to drive his homemade car at speeds over two hundred miles an hour, at which all things assume an "ethereal clarity" (24), and to take his gospel of high speed and inner peace to the L.A. freeways where it can do the most good to all those drones mired in the sludge and dreck of post-urban Southern California. "Oh, the patience of bovine human, requiring no shepherds, no cowboys and no packs of gaunt coyotes to keep them in line. Only concrete abutments" (101).

On the freeway, the narrator has the same illusions about freedom that Huck had on the raft, and his desire to stop moving is roughly equivalent to Huck's ability to drift northward. His car runs on pure alcohol, and comes equipped with peel-back layers of paint to help him dodge state troopers—as well as a few other design idiosyncrasies: "There's no backing up on the freeway. That's rule number one. Not that there are any rules. But when we built the transmission we didn't bother putting in a reverse gear. Who needs the extra clutter? We aim to go ahead" (40–41). The novel itself, appropriately, is one long chapter, its narrative pace averaging just under the magic two hundred mark. But that's why it's so striking to run up against lyrical passages like the narrator's evocation of undone jigsaw puzzles, pieces "scattered like gems through a forest," "precious as antique coins, raining down upon the city like hailstones" (109), or his empathetic reading of the terrified face of a homeless woman covered in a sweater of crushed aluminum cans, who reminds him of the homeless man covered in kelp:

> And she's looking at me that same way: Do I have it right? She doesn't know either, that she isn't the only one who suspects she doesn't. Isn't this how they do it? They stash their wealth and their valuables close to their chest, I've seen them. Don't they? *Don't they!* What is it, oh God what in the world is it. . . . Please can't you see, *I'm* the one who's scared. (61)

Shortly thereafter, D-9 Cats bulldoze the homeless encampments into the ground, and the bulldozers appear again in a surreal scene at the end of the novel, clearing a freeway jammed with newly abandoned cars whose owners have squeezed out of their windows just in time to avoid being compacted. Meanwhile, with the unlooked-for aid of a disabled artist-sculptress, the narrator fulfills his *other* mission, finding his twin sons in Orange County and instilling in them, if no one else, his love of freedom and velocity. The narrator rides off at two-hundred-plus next to an oddly similar alcohol-burning car whose driver is unknown to us, but surely we will be hearing from Bayard Johnson again before too long.

The rest of Avant-Pop promises more than it delivers, but I'm beginning to suspect that's the nature of the beast. Mark Amerika's hyperactive hypertext, *The Kafka Chronicles* (1993), jumbles journal entries, Gregor Samsa, sex, drugs, the adventures of Blue Sky and Alkaloid Boy, and the CNN ratings coup known to Amerika as "Amerika at War: The Mini-Series." Interesting and at times genuinely corrosive, but only in Donald Wildmon's America could sex and violence carry the weight of political subversion they're laden with here. (Of course, we may wind up living in Donald Wildmon's America, in which case Mark Amerika should become required reading.) Philip Lewis's *Life of Death* (1993) dissipates its initial energy in a similar fashion, by losing it in scene after scene of adolescent-anarchic sex and relentless caricature. The narrator, Louie Phillips (cough), opens with caustic wit and some politically pointed signifying on the signs of the times: his father's "Third World" bookstore fails when he's outhustled "by those other book-runners next door to him, young white hippies and ganja-smokers who, apart from selling Gaddafi's Green Book, took heed and stocked *The Closing of the American Mind*" (11–12). Living in a dystopian-multicultural suburb of D.C. composed of warring "Africans and Indians and Jamaicans and Nicaraguans and Vietnamese and Cambodians" (33), Phillips leaves home, drops out of school, and takes a typical Reagan-era job at the Dummheit Café, a mall restaurant owned by a multinational up to its ears in South Africa. Here he learns "what they mean when they say 'equal opportunities' these days—everyone's got an equal opportunity to get his ass kicked high and wide" (25). A strong beginning, and an accurate sense of labor in the post-Fordist economy, too. By page 50, however, Lewis's satire gets repetitive, which means it stops being satire. But then, it *is* his first book.

Eurudice's *F/32* (1990) tells the story of Ela's vagina, which is stolen from her not long after it is severed from her body by a blind black man who removes it with a knife on upper Fifth Avenue while a curious crowd looks on. Apparently someone swiped the jar Ela was carrying it in after it was severed, and replaced it with a 35 mm wide-angle lens. Well, that explains the book's title: the lens "shuts at aperture f/32, the smallest opening in any lens" (133). Soon Ela is searching for her cunt (the book calls it this a few thousand times) by way of the *Village Voice* personals. The thief, meanwhile, tries to seduce the cunt by playing it some "sexy watery music—Jarre's *Equinox*" (155), but it escapes, only to appear later as a guest on late-night talk shows. As in Pynchon's *V.,* characters search frantically for a character known as V who's apparently responsible for all kinds of random destruction; this V, however, just happens to be a cunt. It's about this time that killer female genitalia terrorize the New York metropolitan area. Ela and her cunt are eventually reunited (I will refrain from puns about wholes), and the novel closes with a first-person excursus on sex and mirrors: "in the beginning of all there was the mirror, on every side, wide open. . . . [F]rom my safe mirror view, this and the other side were identical" (273). Gogol meets Lacan, and the two tiptoe through the tulips, speaking together. Eurud-

123

ice's *F/32* embodies the best and worst of FC2: annoyingly outrageous, hyperliterate, funny, theoretically sharp, and really revolting. Sometimes all on one page.

Much of the same can be said of Derek Pell's contribution to the collection titled *Avant-Pop,* an S/M version of Strunk and White's *Elements of Style* ("by the Marquis de Sade, with Revisions, an Introduction, and a Chapter on Writhing") that starts from Roland Barthes's claim that Sade's "pornographic messages are embodied in sentences so pure they might be used as grammatical models." Ricardo Cortez Cruz's *Straight outta Compton* (1992) also contains its share of anything-goes sex-n-violence, coded not by de Sade or E. B. White but (as the title suggests) by gangsta rap à la N.W.A. Much of the novel reads as if Cruz is to Ishmael Reed what Reed is to Richard Wright; characters' dialogue turns out to be samples from *She's Gotta Have It, Star Trek,* and funkster Morris Day, and at one point Reed himself is signified upon: "while it was dark outside," writes Cruz, "Rachel practiced hoodoo—whatever that is" (55–56). Rodney King figures prominently; Chris Rock's *New Jack City* crackhead can be spotted in the wings; and old-school rap giant Doug E. Fresh puts in a cameo as well, and gets his shoes stepped on by boyz n the hood. The novel as a whole reads something like naked lunch in South Central, which is to say (among other things), incoherent, powerful, and savvy about a world Burroughs didn't much acknowledge.

Overall, Avant-Popsters generally require of their readers more hybrid cultural literacies than do their neo-modernist cohorts, as well as a greater tolerance of bad taste, anarchist political farce, and sexually explicit shenanigans. To this last item, a feminist neo-mod like Lou Robinson might respond, "Why does everybody talk about the body? All the best states leave it behind. It is a launching pad, a jetty" (*Napoleon's Mare,* 3). Or, to put the distinction more simply, one might say A-P characteristically appeals to a younger readership than the rest of FC2's list; but since I myself, by accident of a 1961 birth, am either the youngest sibling of the boomers or the oldest specimen of Generation X, my demographic loyalties are somewhat divided on this score. I'll take Bayard Johnson's narrative euphoria, for instance, over the crafty but deliberately dreary sentences of Yuriy Tarnawsky, whose *Three Blondes and Death* (1993) declares that it is written according to an intricate mathematical formula devised by its author. That formula escapes me, though I did manage to notice that the novel's four-hundred-odd pages are filled with short, affectless sentences none of which contain any punctuation other than periods. When the principal character, the unpronounceable Hwbrgdtse, rapes a girl in chapter 47 of part 4, the narrative effect of Tarnawsky's sentences is downright terrifying. Most of the rest of the time, though, they seem . . . well, methodical, as if they were generated by a mathematical formula devised by the author. Where Avant-Pop smells like teen spirit, *Three Blondes and Death* sounds like late Beckett without the cadences.

Narrative experiments every bit as forbidding yet (on the whole) less clinical can be found in Lou Robinson's *Napoleon's Mare* (1991), D. N. Stuefloten's *Maya* (1992), and Rosaire Appel's *transiT* (1993). Robinson's novel is the least narrative of the three, and it's published along with some of her "prose poems" that don't look all that different from the "novel" they follow. Her protagonist "makes balance" by practicing a kind of cut-up technique, collecting and juxtaposing words and pictures culled from various magazines: "a giraffe next to a sandstone arch" (5), "a spitting bobcat next to Bernadette Devlin facing it with a glistening snarl" (12). If the clarity and asequentiality of the images suggests echoes of Mark Leyner, the book as a whole registers a much deeper debt to Gertrude Stein and Djuna Barnes—as when, in the section entitled "A Lesbian Is a Memoir," we read, "A lesbian is a woman who reads without respecting anything. Where is the authority in these words? A lesbian or a memoir neither has nor answers to authority. . . . Who can say. Who can say what? Who can say what a lesbian is. A memoir is a manifesto is a lesbian fucking literature" (21–22). Now, let's see what T. S. Eliot would make of *that*.

Rosaire Appel's *transiT* swirls around a series of overheard conversations or half-glimpsed, ill-understood scenes, developing a kind of vorticist narrative whose vortex is a café we enter again and again at different times, from different angles. Sourceless italicized passages appear—and then reappear in the mouths of unnamed characters; men and women have crises and converse intensely in hot, claustrophobic rooms; someone misses the last train; someone climbs a hill, parched with thirst; someone else dreams of climbing a hill, parched with thirst. The café conversation returns, different snippets overheard this time. Threading through it all are Appel's voices, some flat and toneless, others of quite distinctive pitch and timbre: "I turn and look out the window. Because of the smoke it feels like dusk, all afternoons are like this. The sky injected with the fat of our lives turns sullen and bloated, a stench comes with it" (37). The overall effect is less like "applying the ideas of painting to fiction," as the author's bio blurb suggests, and more like dreaming in Robbe-Grillet.

D. N. Stuefloten's *Maya* begins with a literal version of Dos Passos's "camera eye," recording by way of stage-direction prose the filming of three actors, before proceeding into a ferocious vision of film, rape, and "theaters" of war. "Maya" could have something to do with ancient peoples of the Americas, or with the Hindu concept of illusion personified as a woman; and as it happens, Stuefloten hinges much of his depiction of American illusion (in Hollywood) and American militarism (in Central America) on the figure of a woman, Virginia White, an aging American actress who (rumor has it) once entertained the troops in Vietnam nude from the waist down. (Stuefloten's imagining of this unlikely scenario implicitly refers to the "Playboy bunny" scene in *Apocalypse Now,* a film to which much of this book is clearly indebted.) When at last the narrative frames are ruptured—when the American military has obliterated the village where the actors are located, and like Ronald Reagan we are no longer sure what's movie and what's history—the person on whom the weight of Stuefloten's outrage falls is the woman herself, who is repeatedly raped and brutalized by American servicemen as sundry machines from Sikorksy, McDonnell-Douglas, and Grumman drop their ordnance. The epilogue attempts some leaky damage control, when Stuefloten writes of the writing of the novel, and is rebuked by his lover: " 'You are hard on women,' she said. 'I dont think you can claim innocence in this affair. To some degree you must admit Virginia White is your creation' " (135). But since the chapter and novel end with Stuefloten and his lover having sex (despite her initial resistance), reenacting in the process one of the scenes from the "camera eye" section of the book as well, it's clear that the author has the last word, as the novel contains his lover's critique only to "contain" it.[4]

It's not hard to spot the modernist master lurking behind Kathryn Thompson's *Close Your Eyes and Think of Dublin: Portrait of a Girl* (1991). The novel revisits not only Joycean language and narrative pyrotechnics but also some of Joyce's primary obsessions, sex and Catholicism chief among these. Some of Thompson's re-Joycing is done by way of Adrienne Rich, as in "molly she chose life in the great good indisputable Bloomsbook as readily as the deer chose his appointment in the tack room" (190) or "like Hester Prynne and St. Joan I find I am the real women all dressed up in men's words; you've got to get to know the scrolls and nuts of the instrument that defiles you, eats you alive" (41). Such passages call to mind "Snapshots of a Daughter-in-Law," particularly the lines that reply to Yeats's "Leda and the Swan," "A thinking woman sleeps with monsters. / The beak that grips her, she becomes." Molly herself turns up on occasion, once mouthing her famous final words to an abusive creep who bombs abortion clinics in his spare time, and later telling us, "I learned them gibberish words what with beer commercials the conflagration of souls it's what you men talk about chewing with your mouths open behind our backs along the Strand" (189). Thompson herself, when she's not paring her fingernails like the God of creation, gets off a few zingers about Joycean fathers and sons— "Father, son, and holy ghost, sounds like another bachelor party, another gangbang another spring break at Lauderdale where yours truly jumps out of the cake (god's most intelligent creature after man)" (86)—and leaves no doubt why her portrait of a girl can't look much like Joyce's chronicles of Dedalus:

> I wasn't allowed up there, in the sacristy, as a girl, except to vacuum and dust at six a.m. before mass and before I was even *up* and the way the priest had spoken that word *sacristy* I knew it was the place where men played poker and rolled the everlasting die at night. . . . I should like to know if that is where they keep the wives and children, if that is where Hester Prynne was taught the one letter Alphabet. (81–82)

And the reason she's meditating on gender, the sacristy, and power/ knowledge is that sure enough, her alcoholic grandfather has died on Bloomsday.

As Joyce is to Thompson's *Portrait,* so is Kafka to Kenneth Bernard's haunting, strangely evocative *From the District File* (1992). The novel's first half is benign enough, narrated by one of those citizens who was never so rudely awakened as was Joseph K.—one of the "retired and disabled clerks of division two level of the municipal civil services" (20) whose life, like ours, consists of minutely dissected hassles at the post office and supermarket. The occasional disruption—the disappearance of Mr. M., a brutal beating by Grodek—fails to disturb his torpor and equipoise. It is only when he comes of age and joins one of his society's mandatory "burial clubs" that things turn sinister. The clubs are composed of groups of twelve, contracted to meet socially on a regular basis—attendance is taken and reports are filed—and, most of all, arrange and attend each other's funerals (clubs are replenished when their numbers sink lower than four). Our unassuming narrator comes across subversive burial clubs doubling as refuges for "youths who had been expelled from school for laziness, disruption, or rudeness, who had been in trouble with one authority or another, who in some cases had been branded criminals" (83). Like a bad librarian from the land of Borges, he begins to slip undetectable errors into his official accounts of the club, creating "false cross-references" and "meaningless abbreviations, like 'born 1931, *vit.*' or 'retired at sixty, O.B.C.' " (72). But as in Kafka, the undetectable is detectable, as a "minor clerk in district headquarters" says to him over coffee that "she had begun to notice some extremely minor but regular errors in my otherwise perfect reports" (84). The club's annual culture entry, "Herring Gulls at Dawn," attracts notice for its lack of moral and plot, and the club is disbanded and restocked with company goons. Despite the narrator's conviction that he is "no enemy of the state" (112–13), he begins to fear that his new burial club comrades plan to murder him. He soon lights out for the forgotten margins of the metropolis, where the former street performers have been banished, where he lives out his last days in a landscape half Paul Auster's *City of Glass* and half Ray Bradbury's *Fahrenheit 451.*

Rounding off my account of the neo-mods is Jacques Servin, whose eccentric collection of cosmicomics, *Mermaids for Attila* (1991), marks him as the unacknowledged child of Gertrude Stein and Italo Calvino. In "Life (A Porn Story)," the foreman surrounded by bloodhounds sees how it all ends:

> "I am safe," the foreman speaks in the desert, his toe designing arabesques, as the bloodhounds fall from the sky like entertaining confetti. The world is over, all over, all done, and the foreman is safe. The sky rains down its miraculous bloodhounds, replacing the web of information that never stops, and the foreman gazes with perfect solemnity at the clear, smooth expanse of void that he knows is the darkness of space right beyond the ionosphere of his planet, the earth. (18)

Like a good symbolist poet, Servin shows us time and again that his craft can sustain itself on sound alone, regardless of sense and syntax. But like a good Dadaist, he also intersperses his forays into "pure" poetry with visions of the irreality of political history: "the president has recuperated from his violent death by opening dialogues with the worlds of the ancient Irish" (40). That line, of course, comes from the story "Regarding the Uncustomary Events of 1989 Which Occurred Here and There around Northern Europe," and if you don't believe it, you can ask Hitler, who appears in one story shouting, "Placebo! . . . The ungulate memorize! The fecund 'isms' of my lunacy growl up the spiny decorum of the alma mater" (36). With a little reshuffling of the file cabinet Servin could conceivably find himself listed under Avant-Pop, next to whimsical musical duo They Might Be Giants.

My personal favorites from FC2, though, are the unclassifieds, Omar Castañeda, Cris Mazza, and especially Richard Grossman. Castañeda's *Remembering to Say "Mouth" or "Face"* (1993) is a motley assortment of stories, ranging from realism to surrealism to magic realism to just plain magic. The title refers to the narrator of the first story, "On the Way Out," as he begins to recover from a coma, his body "a coil of shattered pieces each screaming out in its own anguish, each begging for relief" (27). "I Tell You This" does indeed tell you this, narrating its own composition until "this final paragraph, this vertiginous whorl that finally stops at a very curious 'you' " (36). "Crossing the Border" tells a harrowing tale of a trip from rural Guatemala to *el norte,* initiated by a dead man, his concerned friends and relatives, and the traveling shyster who insists that the corpse be buried in To-Bend-Word, a/k/a Klouwer Town, somewhere in the northern United States. Upon crossing the border, the *As I Lay Dying* journey adopts a new currency, as the travelers are set upon, sexually assaulted, and shot by seven young drunken Texan men.

The border crossings in Castañeda's stories turn out to be still more challenging than this: after "Crossing the Border," the collection follows with a section called "Remembering," consisting of six stories derived from the Mayan *Popol Vuh,* the sacred book of the Quiché Indians, which was destroyed in the sixteenth century by Spanish conquistador Pedro de Alvarado. The stories tell various creation myths and birth narratives, chiefly centered on characters such as Blowgunner and Jaguar Sun, twin sons of Blood Girl, who is banished from Xibalbá after being impregnated by the fruit of a tree. What's particularly odd about this full-scale reproduction of Quiché tales from the *Popol Vuh* is that the story before "Crossing the Borders" narrated a very different border crossing, a sketch of a writer and his anthropologist lover, whose relationship does not survive the tension created when he receives a fellowship to study the Guatemalan Maya and she doesn't—despite the fact that the Indians of Guatemala are *her* specialty, whereas he knows nothing about them whatsoever. As if his *Popol Vuh* stories weren't already alien enough, in other words, the Guatemalan-born Castañeda is out to make sure we don't assume too readily that the border is more permeable North to South than it is for his ill-fated travelers trying to move South to North.

Cris Mazza's third FC2 book, *Revelation Countdown* (1993), is among the Black Ice series, but Mazza herself—author of *Animal Acts* (1988) and *Is It Sexual Harassment Yet?* (1991)—looks somewhat out of place among the slash-and-burn fictions of Avant-Pop. She's an accomplished, deadpan portraitist, whether she's detailing the private bodily dissipation rituals of a professional photographer driving aimlessly cross-country after a big book-signing tour (in "Revelation Countdown"), sketching the psyche of an abducted young woman who commandeers and steals her abductor's pickup truck (in "Guys with Trucks in Texas and California"), or rendering the brief romance between a male senior editor and a female assembly-line worker in a romance novel factory (in "Wistfully"). In a different decade Mazza's ear for hopelessness and her muted prose style would have pegged her as a minimalist; in *this* decade she plays quite another role, unwillingly testifying to a phenomenon of which Roland Barthes could not have dreamed—the Death of the Midlist Author. Yet she's capable of a few roller-coasters of her own, as in the title story of *Is It Sexual Harassment Yet?* whose dual, side-by-side columns tell two versions of a restaurant's "hostile environment" that are so drastically at odds (and each so internally coherent and convincing) as to make Clarence Thomas and Anita Hill look like Roy Rogers and Dale Evans. The closing story in *Revelation Countdown,* "Between Signs," sums up the collection's concerns by juxtaposing vignettes of a couple driving through the Southwest in a mutual sexual frenzy to what appear to be banal, crass, or meaningless road signs like "Make a Bee-Line to ROCK CITY" and "Gospel Harmony House Christian Dinner Theatre." You're encouraged, in short, to read between the signs, where the intimate details of "private" experience seem to carry the spiritual significance and human import lacking in the "public" announcements of truckstops, roadside attractions, and dangerous curves. It turns out, however, that a few of those signs are of greater human and social significance than anyone's sexual desire for anyone, such as

WARNING:
THIS ROAD CROSSES A U.S.
AIR FORCE BOMBING RANGE
FOR THE NEXT 12 MILES DANGEROUS
OBJECTS MAY DROP FROM AIRCRAFT

—or "STATE PRISON Do Not Stop for Hitchhikers." Once you've seen a few signs like these, you look again at "Ice Cream, Divinity, Gas, Picnic Supplies / Real Indians Performing Ancient Rites," and you find that Mazza has managed to deconstruct private and public, banal and uncommon, almost without signing her intention at all.

Richard Grossman's novel *The Alphabet Man* is nothing less than a psychotic and singularly moving tour de force. Its narrator/protagonist is Clyde Wayne Franklin, famous parricide and foremost poet in the United States. If you can imagine a world where a crazier Charles Baudelaire and a milder Charles Manson inhabit the same body, you're on the right track; and if I tell you that Franklin is the unwitting pawn in an elaborate (Pynchon–meets–Oliver Stone) plot to kill the likely Democratic nominee for president, a wildly popular anti-Pentagon lefty (that's the tip-off that you're dealing with *fiction* here), I'm really not giving away much of what makes Grossman's novel worth getting lost in. (Though it should be noted that Franklin is chosen as the assassin not only for his psychological instability but also for his name, which fits the three-bland-name pattern of Lee Harvey Oswald, James Earl Ray, and Mark David Chapman.) Franklin's introduction, "America loves a murder, and I am a murderous American" (17), isn't great; it's certainly not as inspired as the opening-and-refrain of Robert Steiner's *Broadway Melody of 1999* (1993), "Munchkins littered the yellow brick road" (7). Indeed, *The Alphabet Man* initially promises to be a morning talk show gorefest, or at best a cross between the demeanor of Dashiell Hammett (as Franklin turns gumshoe, tracking down his erstwhile prostitute lover) and the morbid fascinations of Thomas Harris (whose Hannibal Lecter novels have given us the films *Manhunter* and *Silence of the Lambs*). *The Alphabet Man* looks, from this vantage, like standard 1990s fare from the police blotters of the blood-soaked U.S. of A.: just another American psycho. Besides, building a novel on the "confessions" of criminals is an idea that went out with Daniel Defoe.

But over the course of four hundred typographically peculiar pages, Grossman weaves together a luminous, deeply lyrical story out of the unlikeliest assortment of cast-off materials and popcult trash—not just old detective novels, but also the kinds of stories of child abuse, multiple personalities, and psychotic clowns that you find more readily among the accounts of Elvis sightings than in the pages of "quality" fiction. Much of the novel is narrated by the voices that live inside Franklin, as Grossman blends Bobcat Goldthwaite, Ring Lardner, and Mister Bones from John Berryman's *Dream Songs* into a heteroglossia as dazzling as it is maniacal:

I know that you got it all tied up, but Jeez Al (don say dat word)
I said Jeez Al (I say don say it or else) I said OK I won't say,
now look fella that Clyde guy he treats me like his chicks man
and he's got me all penned up out here in the Encino of his
head, I mean the man is one sick motherfucker, and I have to
follow this grisly plot of his and stick in my four bits, and he
just goes off, he keeps his nose on the ground plowing through
the crap and then he just flies, he goes off and I've got to pick
up the pieces. (115)

The novel closes with one of Franklin's poems, a 170-line lament called
"Barathrum"; as Franklin's friend and editor L. Jerome Dalls notes, "a
barathrum, from the Attic Greek, was an ancient pit in which con-
demned prisoners were thrown, and hence has come to mean the depths
of hell" (431). The poem is remarkable not only in its own right but also
because it pulls off the unlikely feat of concluding the novel in the mode
of threnody:

> *Oseh shalom bi'm'romav, hu ya'aseh shalom aleynu v'al kol yisrael, v'imru*
> *amen*
> these pits are cicatrices of earth over deep peopled wounds we shall
> never recover the scars are ours forever and ever. *Amen.*

The poem in turn refers us back to one of the novel's epigraphs: "A
Hassidic tale recounts the story of an illiterate man who could not read
the prayers for Yom Kippur. The man left the synagogue and walked out
to an open field to recite the letters of the alphabet, one by one, asking
God to put them together to form a prayer. God received the man's
prayer and was pleased." Not least among Grossman's many achievements in *The Alphabet Man* lies in the poignancy of how the novel itself
becomes Clyde Wayne Franklin's prayer of atonement, a prayer whose
alcoholic, schizophrenic vessel winds up being one of the most memora-
ble characters I've encountered among the American fictions of this
century.

It will doubtless be clear by this point that there is no single thread
running through the recent work published by FC2; on the whole, this
work is generally more shocking, more difficult, or more iconoclastic

133

STRAIGHT OUTTA NORMAL

than your average offering from Knopf, Morrow, or the Congressional Budget Office, but it's impossible to chart FC2's cultural location in the pseudomilitary terms of *avant* and *garde*. Nor is it possible to predict where you'll hear of this work next: it was, for instance, the lunatic multiple-streams-of-consciousness of *The Alphabet Man* that got a friendly notice from *Details* magazine, the high-gloss bible of fashion-conscious twentysomethings, and Ricardo Cortez Cruz's hip hip-hop hype, *Straight outta Compton,* that got props from the *Nation,* the no-gloss paper of record for unfashionable lefties.

But I can predict one thing safely: whatever happens to the NEA, ten or twenty years from now, *Alphabet Man* and *Straight outta Compton* will still be around. As large houses cut more and more unprofitable midlist writers from their rolls, FC2 may pick up more writers like Cris Mazza and Kenneth Bernard, whose writing, in a less profit-driven fiction market, would certainly be accessible enough for a few of the smaller conglomerates. And FC2 will always be the place for utterly unmarket-able or extremely abrasive work, such as Samuel Delany's *Hogg* (1995), which can make some of the grisliest scenes from *Naked Lunch* look like teatime with Jane Austen. Yet one truly crucial thing about FC2's publishing enterprise is usually overlooked: like many small presses and academic publishers, the collective does not take works out of print if they're not "selling." Three hundred copies left in the warehouse? Fifteen sales last year? Doesn't matter, the book stays available. And as a result, you can still get your hands on Clarence Major's and Peter Spielberg's novels from the 1970s even though the contemporary work of less experimental writers is no longer for sale. What's more, FC2's backlist will likely remain of greater interest than will that of most university presses: it may not matter much to the world if a twenty-year-old study of nature imagery in Faulkner remains somewhere in the remnant bin for only three dollars, but it's important that *Reflex and Bone Structure* is still at your fingertips, just as it will be important in 2020 to be able to get *The Alphabet Man, Damned Right,* or *Napoleon's Mare* simply by filling out an order form.

This, then, is the most important thing about the press, to my mind: whatever else it does, Fiction Collective 2 performs the critical task of sustaining this nation's weirder literary heritages against the logic of Time Warner and the Sears Financial Network, which will foster "novelty" only so long as it is quickly succeeded by more "novelty." FC2's novelty, by contrast, is here to stay. Among the many functions undertaken by FC2 as a nonprofit "cutting-edge" press, it is of signal and lasting consequence that the collective be a literally *conservative* force for the preservation and transmission of Avant-Pops, neo-mods, and unclassifieds everywhere.

EPILOGUE: FC2, THE DALKEY ARCHIVE, AND THE UNIT FOR CONTEMPORARY LITERATURE

One fine April day in 1994, again at the invitation of Charles Harris, I drove an hour up the interstate to the new Unit for Contemporary Literature at Illinois State University in Bloomington-Normal, Illinois, where Richard Grossman would be reading from *The Alphabet Man*. The reading took place in the studio of noted painter and sculptor Nicholas Africano; the reception, during which Grossman handed me a card that said "Dick" and nothing else, was held in a building called the Orphanage across the street. Actually, I learned, until 1978 the building *was* an orphanage: Africano recently bought up the ten outlying buildings that were its residence halls, and his own studio and living quarters were once the recreation center or something. The whole place had that same weird former-institution atmosphere you used to sense in converted Soho lofts fifteen or twenty years ago (only sadder — I mean, *orphans,* for Chrissake, many buildings of them). Attendance was great: the studio was full, plenty of post-literate folks, interesting hair configurations, great design sense, sparse, stylish furniture, chain smoking, and a couple of Kathy Acker look-alikes thrown in for good measure. And if I were still the parochial New York expatriate I was in 1983 when I ventured out of the New York City metropolitan area for the first time, I would have wasted a good portion of the evening thinking to myself, *what is all this doing in Normal, Illinois?*

The Unit for Contemporary Literature began its programs in February 1994; it's mainly an umbrella group for FC2 and its sister press, the Dalkey Archive, but its other current projects include researching and publishing a study of the history of small American literary presses. And then when Ron Sukenick folded the University of Colorado (Boulder) branch of FC2, Normal also picked up publication of the *American Book Review,* one of the nation's liveliest general-purpose reader's guides to everything.

All this from a place that also publishes periodicals such as the *Spoon River Review,* the *Review of Contemporary Fiction* (via the Dalkey Archive), *Mediations* (journal of the Marxist Literary Group), and the *Exquisite Corpse,* a quirky pamphlet-sized writers' forum for prose poems, disjecta, and curios (the first copy of the *Corpse* I picked up pointed out the uncanny parallels between *Love Boat* and *Star Trek: The Next Generation*—Gopher/Data, Troi/Julie, Doc/Riker, and the bald guy who adjusts his uniform when he's flustered—and it did so many months before Time Warner's *Entertainment Weekly* pointed out the same parallels in less hallucinatory prose). Almost all these journals started elsewhere; *Exquisite Corpse* may have covered more miles than your average chain letter. But there are apparently many roads to Normal, and if my visit (on only one of those roads) is any gauge of things to come, Normal may be on its way to becoming the alternative press capital of the United States.

Having perused the FC2 list, I returned to Normal in 1995 to learn a few things about the Dalkey Archive Press.[5] Curiously, FC2 and the Dalkey Archive have gradually become something like complementary contrasts. Where FC2 solicits the as-yet-unheard-of writers of the 1990s and 2000s (whatever we'll call that decade), the Dalkey Archive has been retrieving valuable writers of the 1960s, 1970s, and 1980s from the remainder bin with an energy that provokes both admiration (Gilbert Sorrentino! Paul West!) and alarm (John Barth's *Letters* is out of print?). In other words, as worthwhile literature drops out of print and out of sight, the Dalkey Archive steps in and preserves it cryogenically. It also publishes promising contemporary writers like Aurelie Sheehan, author of *Jack Kerouac Is Pregnant* (1994), or Carole Maso, whose densely lyrical *AVA* (1993) and *The American Woman in the Chinese Hat* (1994) have won well-deserved rave reviews and the usual small-but-devoted following—though not devoted enough to keep her original publisher, North Point Press, in business. That too, as I've noted above, is very much a sign of the times: North Point, a small commercial publisher, reportedly needed 7,500 sales of each book to stay afloat, and even though it did come up with a few best-sellers, it couldn't manage to snag the requisite .00003 of the American public on a regular basis.

For the most part, though, the Dalkey Archive, unlike FC2, works the side of the street populated by great dead writers down on their luck and talented living writers who've been cut from the backlist: in 1994, for instance, it reprinted Ford Madox Ford's *The March of Literature,* and in 1995 it followed this feat by reprinting Gertrude Stein's notoriously vexing magnum opus, *The Making of Americans,* all 962 pages of it. Together with its parent-and-affiliate publication, the *Review of Contemporary Fiction,* the Dalkey Archive has practically become the country's Djuna Barnes headquarters: the *Review* did a special issue on Barnes in 1993, and two years later the Dalkey Archive brought out a definitive edition of her 1936 masterpiece, *Nightwood*—including all the material T. S. Eliot and Barnes's friend Emily Coleman cut from the original manuscript when it couldn't find a publisher, along with Barnes's drafts for the novel and an introduction by Penn State assistant professor Cheryl Plumb. In 1990, likewise, the Dalkey Archive had republished Barnes's novel *Ryder*—although that text, eviscerated on publication in 1928, cannot be restored as can *Nightwood.* As Paul West's afterword explains, once the book was censored by the New York Post Office, "Djuna Barnes and Charles Friede, an editor from Liveright, sat there in Paris removing passages having to do with body fluids" (243); *which* body fluids particularly PO'd the PO we'll never know, but I'll bet that Representative Peter Hoekstra and his friends don't like them either.

Barnes, Stein, Ford, and Company: it sounds like a modernist investment group—and it is, in a way. Even the Dalkey Archive's new writers, like Lauren Fairbanks, author of *Sister Carrie* (1993), or Christopher Sorrentino of the Flying Sorrentinos, author of *Sound on Sound* (1995), have roughly the same relation to high modernism as FC2 writers like Kenneth Bernard, Kathryn Thompson, or Jacques Servin. Whether working in the tracks of Kafka, Woolf, Joyce, Beckett, or Breton, these people do what modernist fans used to call *writerly* writing—making readers work for their money, chronicling the relations between sound and sense, giving language a well-deserved holiday or putting it on the double shift. The result is either modernism made new again or a post-postmodern profusion: but together, the presses of Normal are producing almost three dozen books every year, many of which either revive modernist classics, pay homage to modernist legacies, or revisit and critique modernism from cultural positions all over the map.

In a sense, though, that debt to modernism isn't surprising. The Dalkey Archive's first reprint, back in 1984, was Gilbert Sorrentino's *Splendide-Hôtel,* which had just been dropped by one of modernism's initial clearing houses, New Directions Press: the pass of the baton could hardly be clearer. But according to DAP founder John O'Brien, becoming a new NDP wasn't in the original game plan at all; his half of the Normal consortium, unlike the Fiction Collective, started out with no specific agenda or long-term goals. In fact, the Dalkey Archive grew out of the *Review for Contemporary Fiction* (and not the other way around), which began in 1981 simply as an effort to put some contemporary and postwar writers on the table for discussion—Hubert Selby, Paul Bowles, Julio Cortázar, Jean Rhys, Ishmael Reed—whether or not anyone paid attention. "I figured we'd run five years or so, sell maybe 150 copies, and then vanish," says O'Brien. "Then we'd become the kind of journal about which people say, 'whatever happened to *that?* "[6] But initial sales surpassed these humble expectations, and O'Brien, faced with a surplus after only three years, decided to take a chance and reprint *Splendide-Hôtel;* two years later, on a similar lark, he took a new book, Paul Metcalf's collection of short stories *Where Do You Put the Horse.* "Again, I figured 'just this once'—and it was our only book for 1986," O'Brien told me. Nine years later, the Dalkey Archive houses no fewer than 168 titles, including treasures like Felipe Alfau's *Chromos* (1990), Viktor Shklovsky's *Theory of Prose* (1991), and, of course, Flann O'Brien's *The Dalkey Archive* (1993). Now *that's* entertainment.

Few cultural institutions can say so much for themselves; fewer still have done so much in so little time. And yet the prospects for DAP,

FC2, UCL, and ABR at ISU are thoroughly mixed; the dominant mood of the place runs from pride and joy to cautious pessimism. In part that's because their writers and their enterprises are so wildly various, the writers themselves ranging from the renowned to the unknown, the brilliant to the boring. But it's also because the cultural weather is so volatile: both enterprises are very heavily dependent on the continuing kindness of strangers, and though many strangers have been kind indeed, O'Brien admits that the Dalkey Archive simply would not exist if not for the National Endowment for the Arts and granting agencies like the Lila Wallace-Reader's Digest Fund. "The *Review* might be able to make it," O'Brien says, adding, "just barely. But the press wouldn't stand a chance."

I spoke with O'Brien in 1995, and his words now sound chilling to me in 1997, as FC2 has become the stick with which Peter Hoekstra is trying to flog the NEA. On the other hand, as I think back to Richard Grossman's reading and that eerie feeling of doing the wine and cheese routine on the grounds where scores of Dickensian waifs once timidly held out their bowls for gruel, I wonder whether there isn't some reason for hope. Despite the shrinking of the imprints, the intransigence of Republicans, and the question of whether Normal, Illinois, can keep up all this manic publishing activity—despite all this, there's something piquant, even encouraging, about the thought of orphanages being converted into Units for Contemporary Literature somewhere in the heart of the heart of the country: few cultural developments, I think, could run more directly contrary to the *zeitgeist* of the Age of Newt. Perhaps, indeed, *that* is the real reason the nonprofit presses of Normal now find themselves under assault.

NOTES

1. Individual Fiction Collective writers have, however, fared better in academic criticism than the collective as a whole. Sukenick and Federman are often cited alongside Coover and Pynchon as 1970s postmodernists; Major and Dixon are sometimes mentioned in the same breath as Ishmael Reed. To date, however, I have only found one article devoted to the Fiction Collective *qua* Fiction Collective (see Quartermain).

2. *Maximum RocknRoll* publishes monthly. Its circulation figures are hard to come by, since its distribution is so broad, but I do want to point out that its "street" credibility is unquestioned: any alternative noise–producing musician will tell you it's the most authoritative review in the business. Harald Hartmann of *MRR* reviewed two of the first Black Ice books—Mark Amerika's *Kafka Chronicles* and Cris Mazza's *Revelation Countdown*—and gave them both a very rare *MRR* thumbs-up. In other words, Black Ice's affiliation with academe was not a barrier to *MRR,* and that really is saying something.
3. Telephone interview with Sukenick, 18 February 1994.
4. This is the only politically correct point I'll allow myself in the body of this essay. For, after all, some FC2 stuff is seriously offensive not only to literary but also to "liberal" sensibilities, as it no doubt should be: Philip Lewis's *Life of Death* is an equal-opportunities slamfest of ethnic slurs, *F/32* is obviously multiply PI despite its witty takes on pornography, *Griever* is highly unflattering to the Chinese, and depictions of sexual mutilation and violation abound throughout the catalogue. My own attitude is that readers who can't stomach this stuff, be they congressmen or choreographers, should turn instead to the Hallmark industry for sweetness and light. There's just one thing. As the parent of a child with Down syndrome, I do wish Eurudice and Kathryn Thompson would go hit the gym and work off their weird aggressions toward "mongoloids" (Thompson) and "pug-faced, red-eyed, excessively salivating, Downs syndrome dwarves" (Eurudice). OK, I feel better now.
5. For a more thorough analysis of the Dalkey Archive, see Barone; for a historical overview of alternative press publishing in the United States, see McLaughlin; for an account of the contemporary crisis in nonprofit and avant-garde publishing, see Harris.
6. Telephone interview with John O'Brien, 11 June 1995.

EMPLOYING ENGLISH

ENGLISH FOR EMPLOYMENT

I hope thus far to have made a few convincing proposals about the relation of contemporary literature to departments of English, and about the importance of maintaining ethical working conditions for graduate students and faculty in the humanities. I am not claiming that these two topics are intimately related; on the contrary, I admit that ethical working conditions for faculty and graduate students need not facilitate the teaching of literature any more than the teaching of literature guarantees ethical working conditions for teachers. And yet there may well be a nontrivial connection between the relative autonomy of the "aesthetic" and the relative autonomy necessary to the functioning of any critical institution in civil society, including universities. In this chapter I want to explore this connection, but I do not want to do so in a mechanical way: I will not argue, for instance, that academic freedom depends on the idea of disinterestedness, whether that idea is tied to aesthetics (as "purposive purposelessness") or to politics (as "neutrality"). And I will definitely not argue that the academic freedom of faculty should depend on whether we read literature "as literature" rather than as evidence about the broader culture that produces and consumes (oh, all right, *writes* and *reads*) literary works. Yet I do want to suggest that there is something necessarily indeterminate about the work we do in English, and that this indeterminacy may be not only useful to the discipline (in a social-pragmatic sense) but also constitutive of the very means by which we address our subject matter, regardless of whether that subject matter is Shakespeare or *The Simpsons.*

At the same time, I do not want to make a fetish of indeterminacy. As a writer on the independent Left, I cannot *not* be interested in the question of whether the knowledges produced in my discipline might be of any use at all in fostering a critical culture in which progressive political change is thinkable (and therefore possible). I argued in *Public Access* that the institution of academic literary criticism is constitutionally engaged in the project of enhancing what I call "advanced literacy," that is, the training of students in the possibilities and varieties of interpretation; the question continually before us, then, is whether training in the subtleties of interpretation can serve the purposes of democratic renewal. Is advanced literacy as necessary to a democratic polity as is basic literacy, and does advanced literacy have anything to do with the cause of social justice?

The discipline of English is perhaps one of the most vexing and problematic places from which to ask this question. In his foreword to Jerry Herron's strangely neglected but very engaging book, *Universities and the Myth of Cultural Decline,* Gerald Graff summarizes one of Herron's central arguments by writing,

> It is interesting to look at the way the official publicity of humanities departments tends to hedge on this question. English departments, for example, characteristically tell students that if

they major in English they will learn to become independent thinkers who will fearlessly question the established beliefs and institutions of their society—but that, additionally, they will acquire the skills that will qualify them for the professions. If you study English, you will learn how to see through corporate capitalism while qualifying for a job at IBM! (5)

Here Graff captures nicely what I call the "reversibility" of the knowledges we circulate in the humanities, a reversibility that not only serves us well when it comes to training students (whereby we try to come to terms with the fact that advanced literacy can be used for progressive or reactionary ends), but also seems somehow bound up with the specific material of our field, since for the past few centuries, "literature" and "the aesthetic" have been understood to exceed any determinate political use. In other words, just as a major in English can teach you to see through corporate capitalism while working for IBM, so too, somehow, can the study of Shakespeare serve revolutionary and conservative social forces at the same time.

I do not see any way of making Shakespeare—or literary study—into a unidirectional vehicle for political change; I do see numerous ways textual study can make students more critical (and self-critical) readers of the texts that compose their world. I said something to this effect at the close of *Public Access,* anticipating, at the time, that my conclusion would sound too wishy-washy and defeatist for those of my colleagues on the Left who speak as if they had the formula for ensuring that their students would never again read or act in a reactionary way. Curiously, however, the closing pages of *Public Access* were promptly fileted not by the Maoist Literary Group Standing Committee on Counterrevolutionary Criticism, but by the nimblest sushi chef in the business, Stanley Fish—who, in his most recent book, *Professional Correctness: Literary Studies and Political Change,* took issue with my work not for its modesty but for its "hubris" (125). Apparently, what makes me so inviting a target is my belief that interpretations can matter to the world, a belief that runs afoul of Fish's claim (and we shall see whether it is anything more than a claim) that criticism and theory, like poetry, make nothing happen. "What Bérubé never seems to realize," writes Fish, "is that although, as he puts it, 'the "textual" or the "discursive" is . . . a crucial site of social contestation' (264), the people who *study* that site are not crucial players in the contest" (124).

It is hard to see how Fish can maintain this position, since among the people who "study" the site of the discursive are people with the power to write laws, allocate resources, and distribute social and material goods; indeed, Fish himself would not be able to travel so freely between the realms of law and literature if there were not interpretive protocols common to both. Interestingly, Fish proceeds from this point to credit me with recognizing my own "immodest" tendency to overreach—adding quickly, however, that my moment of self-recognition is as evanescent as a blush:

> For a moment, in the closing pages of his book, Bérubé seems close to seeing this [i.e., his point about people who study the site of the discursive] as he retreats from his larger claims and imagines a relatively humble role for academic literary work. "The work of literary critics," he acknowledges, "just is the work of interpretation, and the teaching and training of literary critics is the teaching of and training in varieties and possibilities of interpretation" (263). This is so reasonable and mild that we might think we were hearing the voice of Frank Kermode when he describes our task as creating readers "who will want to join us as people who speak with the past and know something of reading as an art to be mastered" and rejects as "immodest" the notion that by teaching such persons to read "we are improving them, ethically or civilly" (*An Appetite for Poetry,* Cambridge, Mass., 1989, 58–56). But immodesty, along with the hope and the claim, immediately returns when Bérubé lists the rewards awaiting those who enroll in his classes: "We make the promise that if you do these things, if you practice the fine arts of textual interpretation, you will 'get more out of' your readings, in terms of your own symbolic economy: you will learn the process of constructing analogies, drawing inferences, making finer and firmer intertextual connections among the texts you've read, and the texts that compose your world" (263).

I had not actually written that passage as an advertisement for my own courses, but now that I think of it, perhaps I should; certainly the passage is too long to fit on a T-shirt or a bumper sticker, and even quoting Fish's citation of it has taken up an ungainly amount of space in this book. But as a matter of fact, I don't actually use such language in advertising my courses—not because I am modest, but because I think this language is so generic, so thoroughly part of the "common sense" of the discipline, that it would not distinguish the goals of my courses from the goals of anyone else's, with the possible exception of Stanley Fish. The "we" in the passage, after all, does not denote "me and my ideological compatriots"; it was, or at least I hoped it would be, capacious enough to include anyone who thinks that literary study might serve any purpose other than the enhancement of literary study.

What Fish takes exception to in this passage is not my claim that literary study helps you "get more out of" your reading; he seems instead to assume that I have proposed interpretive theory as a means to self-improvement. But in fact the only way in which I suggest that literary study "improves" anyone is that it makes people into more careful readers, and I drew on this language of "getting more out of" reading precisely because this is how literary study is commonly spoken of in the culture Fish and I inhabit (and, following Wittgenstein, I think ordinary language is the "site" at which people describe their lives). How often, dear reader and colleague (if colleague you be), has a nonacademic friend said to you, "I really enjoyed [book X] or [film Y], but you probably 'got more out of it' than I did"? Among people who know better than to think that English professors scour other people's utterances for grammatical missteps, this is the most widely understood function of interpretive study: it helps you notice textual detail, read for subplots, place texts in generic or historical contexts, or make something coherent out of seemingly random features of a text. It leads you, in a phrase, to get more out of what you read; and in recent decades, as the procedures of literary criticism have been applied increasingly to ostensibly "nonliterary" phenomena, just as theorists have denied that there is in principle any difference between the interpretation of signs inside and outside literary texts, the discipline of English has begun to make the promise, implicitly and explicitly, that if you can learn to "get more out of" your reading of Jane Austen, you will also learn to get more out of your reading of other texts, discourses, and rhetorics, be they magazine articles, conversations, Supreme Court decisions, rhetorics of empire, or books by Stanley Fish.

And there is every reason for people to become better readers of books by Stanley Fish; Fish is an entertaining writer who manages, at the conclusion of his treatment of *Public Access,* to nail me down almost perfectly. After citing the passage in which I write of "making finer and firmer intertextual connections among the texts you've read, and the texts that compose your world," Fish summarizes my position as fairly as anyone has done to date:

> Up until that last clause this is a thoroughly conventional, even Keatsian-Arnoldian, celebration of the aesthetic sensibility and its pleasures; but with "and the texts that compose your world" another "promise" is made, at least implicitly: the promise that if you learn to read in the appropriate manner and teach that way of reading to others, you and they will read the text of the world differently and thereby produce a different, and better, world. (124)

The composite picture Fish draws of me here is that of a kind of Keatsian-Arnoldian Frank Kermode on steroids—a picture, I think, that would positively identify me in any police precinct in the profession. For I *do* believe that there are texts that compose our world (even if I don't think that all the world's a text), and I do think that if you learn a few things about how texts work, you will be able to apply your knowledge not only to manifestos but to minstrel shows, not only to *Lycidas* but to law.[1] In the passage of *Public Access* under scrutiny here, I did tend to emphasize the interpretation of written texts: "Historicizing a text, speaking its silences, making manifest its 'latencies,' reading its rhetorics, interrogating its implicit assumptions or explicit propositions about race or gender or nation or sexuality or 'culture'—this is what we do and what we try to interest our students in doing" (263). But I never claimed that the kinds of reading performed and taught in literary study were so specific to literary study that they could not be learned elsewhere; in fact, if I *had* made such a claim, I would have wound up in the same dilemma Stanley Fish finds himself in (a dilemma I will explore shortly)—unable to account for, and thus compelled to deny or minimize, the influence of interpretive theory on extraliterary phenomena. Rather, my claim was that

> in theory, you can do this [i.e., learn to read the texts of the world] in nearly any field of human endeavor, from astrophysics to sports commentary, but you can probably do it best in those fields that give the widest possible latitude to understanding the formative and "productive" aspects of language, where the interpretation of discourses and rhetorics necessarily involves interpretation of the discursive and nondiscursive work that "discourses and rhetorics" have done in the world. (263)

My claim, in other words, is not that my courses will make you a better person, but that literary study, like many other fields of endeavor, will make you a better interpreter of the world—and that *un*like many fields of endeavor, it will attempt to do so by way of understanding the discursive and nondiscursive work that discourses and rhetorics have done in the world.

Fish's argument against my position, like *Professional Correctness* in general, depends on a network of ancillary arguments about disciplinarity, public justification, antifoundationalism, and the function of "critical thinking" (which, as we shall see, has for Fish no function at all, since he does not believe in its existence). And because I largely agree with Fish with respect to disciplinarity, public justification, and antifoundationalism, I think it is all the more urgent for me to take up his challenge and find some worthy worldly work for "critical thinking" to do.

I will not purchase my argument on the cheap: to wit, I will not argue that because Fish himself is a quasi-public figure outside the academy, or that because my own most recent book, *Life As We Know It,* spoke to numerous audiences outside my field, Fish and I between us have demolished his argument that literary critics cannot travel outside the discipline without ceasing to be literary critics. But I will say that when it comes to discussing what kind of people and what kind of knowledges *can* travel outside the discipline (perhaps to have some effect on the world, or perhaps just to go sightseeing), Fish's argument is characteristically sinuous and elusive. If you can imagine a sushi chef applying his manual talents to three-card monte, you'll have a fair idea of what I'm grappling with.

In his critique of *Public Access,* Fish's position on cultural studies rested largely on the proposition that the academy's objects of study, even when they include "the work that rhetorics and discourses have done in the world," have no need for the academy itself. Even if it is true that nonacademic constituencies are relevant to the disciplines of cultural studies, writes Fish, "and that it is so is, after all, the *thesis* of cultural studies, then it follows that these constituencies don't need our help; and, more importantly, there is no reason for them to help *us* even though we may be the theorists of their activity" (122–23). Fish here insists that the objects of cultural studies are in fact blithely ignorant of cultural studies: "after all, they don't do what they do because they read Stuart Hall or Judith Butler or Dick Hebdige" (123). This may be true for many "subcultural" groups, but then again, it is hardly true of them all: Judith Butler's work has in fact been influential for "queer" political activists; Stuart Hall's work has intersected time and again with the films of Isaac Julien; and Dick Hebdige's influence on British subculture should by now be as well documented as his scholarly work on British subculture (although perhaps that's more obvious to someone of my generation and tastes, someone who still listens to the Gang of Four, than to someone like Fish). And although it is insufferably immodest, I myself hope (if for nothing else in the world) eventually to have some impact on the social meaning of Down syndrome, not merely by being a parent of a child with Down syndrome but by having written about it in a way that brings the concerns of literary theory to bear on the under-standing of human disability. If Fish argues that academics are structur-ally and substantively irrelevant to the extra-academic phenomena they study, I believe that the burden of proof is on him, and that he has not met it.

Fish's position, however, is not so simple as this. At one point in *Professional Correctness,* Fish cites an example of a work of literary criti-cism that tried to have a material effect on the world, and the example is G. Wilson Knight's *The Chariot of Wrath: The Message of John Milton to Democracy at War* (1942). Knight's book, it would seem, was a work not of criticism but of wartime exhortation; according to Fish, "the subtitle telegraphs Knight's intention: he is going to search Milton's poetry for passages and images that will provide comfort and inspiration to a nation beleaguered by evil forces; he is *not* going to claim that the meanings he finds are meanings Milton intended: only that the meanings he finds are helpful to the British people in a moment of present crisis" (66). This, then, argues Fish, is not literary criticism at all "and has had no lasting effect on Milton criticism" (67). Fair enough; I would not want to counterclaim that Lauren Berlant's reading of Hawthorne's "The Custom-House" in *The Anatomy of National Fantasy* was or could have been the crucial text that led to the passage of NAFTA in 1993. But that wasn't the issue on the table. The question before us was not "can books about literature have an impact on national policy?" but "can the knowledges circulated and produced in English have an impact on national policy?" To that question, Fish has a slightly different answer, one that does not depend on the wartime aspirations of G. Wilson Knight:

> I am not saying that specifically literary skills cannot be applied to extra-literary contexts, just that those contexts will be unaffected by the application. The rhetorical analysis of diplomatic communiqués, political statements, legal documents, presidential addresses, advertising, popular culture, television news, billboards, restaurant menus, movie marquees, and almost anything else one can think of has been an industry for a long time, but in almost no case have State Department officials or the members of the judiciary or even the publishers of Harlequin romances changed their way of doing things as a result of having read—if any of them ever did read—a brilliantly intricate deconstruction of their practices. Think about it. You are about to open a new business or introduce a bill in Congress or initiate an advertising campaign, but you pause to ask yourself, "What would the readers of *Diacritics* say?" (90–91)

This is rather witty—but did you watch the red card? Look again: it's one thing to say "in almost no case have State Department officials . . . changed their way of doing things" because of (in this case) deconstruction. That kind of claim, like Fish's claims about Stuart Hall et al., seems contestable on its face, thanks to the advent of critical legal studies, which is influenced by deconstruction and which does seek to change the way law is taught and practiced (as Fish himself well knows). But it's quite another thing to suppose that deconstruction can change things only if policy makers and ad execs read its former house journal, *Diacritics*. If we are tempted here to prove Fish wrong by sending free trial copies of *Diacritics* to the State Department or the advertising offices of J. Walter Thompson, I have to think Fish will have led us down the garden path.

So far, this is the terrain according to *Professional Correctness:* books of literary criticism do not change the world, and even if we apply the techniques of literary criticism to cultural phenomena, those phenomena will turn out not to have any need of our criticism. Similarly, Fish admits that "specifically literary skills" *can* be "applied to extra-literary contexts," but insists nonetheless, and a priori, that "those contexts *will* remain unaffected by the application" (my emphasis). It would seem, then, that the interpretation of literature is not relevant to anything but literature. From here, Fish needs to dispose of the next obvious objection: doesn't it sometimes happen that ideas take hold on influential people, and that those who profess those ideas can therefore be said to have had some impact on the world? If academic humanists themselves are not the unacknowledged legislators of the world, might they not still have some effect on actual legislators? Fish's response to this possibility bears close attention. Citing "the exception that proves the rule," he writes,

In the years of the Reagan Administration a number of government officials had links to a network of Straussians, students and followers of the late Leo Strauss, a political philosopher who strongly attacked what he saw as the corrosive relativism of modern thought and urged a return to the normative thinking of the ancients. Strauss's views or versions of them were alive and well in the persons of William Bennett, Lynne Cheney, Chester Finn, Diane Ravitch, and Quayle's Chief of Staff, William Kristol, and it is at least arguable that these and others close to the administration were at least able to influence its policies especially in matters of education, the arts, and civil rights. It would seem, then, that this was an instance in which intellectuals had a direct impact on the political life of the nation. (96–97)

For a moment, it looks as if Fish has made a serious misstep: having started from the premise that literary studies cannot contribute to political change, he now seems to be arguing that it is rare for intellectuals in *any* discipline to have an impact on the political life of the nation. Unfortunately, the New Right knows better, which is why it has spent the past generation creating think tanks, foundations, institutes, and sinecures for reactionary intellectuals to produce study after study alleging that welfare causes a "culture of dependency"; and to its credit, the New Right has been right to believe that intellectuals can matter to the world, insofar as they can eventually lead a nominally Democratic president to eviscerate social services for the poor, the elderly, and the disabled.[2]

But Fish's claim here is of a piece with his arguments elsewhere in the book: yes, ideas do travel, he occasionally admits, but if they do it is not because of the ideas themselves. As for the Reagan-era Straussians, Fish writes,

> If these men and women were influential it was not because of their teachings and writings but because they managed through non-academic connections to secure positions that gave their teachings and writings a force they would not have had if they had remained in the academy where they would have had to wait for some accidental meeting between their "great thoughts" and the powers that be. Absent such an accident or an appointment to public office only contingently related to those thoughts (government officials don't say, "He wrote a great book on the English novel; let's make him Secretary of Education"), there are no regular routes by which the accomplishments of academics in general and literary academics in particular can be transformed into the currency of politics. (97)

For those of you who have not played three-card monte, this must be somewhat dizzying. At first the argument had to do with the study of literature, and the *reductio* was G. Wilson Knight; then the argument was about the influence of theory, and the *reductio* was the idea of State Department officials reading *Diacritics;* here the argument is about the political influence of intellectuals, and the *reductio* is the hypothetical appointment of Michael McKeon or D. A. Miller as secretary of education. Where Fish had claimed earlier that "in almost no case" has academic work had any influence on the polity, now he is claiming merely that "there are no regular routes" by which this influence can travel.[3]

And just as I cannot testify to the influence of deconstruction by pointing to elected officials' citation of *Diacritics,* so too can I not rebut Fish by pointing to the regular, well-traveled routes by which ideas take on material force. But why should that be necessary? Why do "irregular routes" not suffice? Fish's position would seem to entail two rather different answers: (1) such irregular routes do not exist; and (2) even if they do exist, they (a) do not affect the extraliterary contexts to which they lead, (b) depend on "accidental meetings" between thinkers and powers that be, or (c) irrevocably change the goods that travel on them. For a succinct formulation of alternative (c), we need look only at the opening paragraph of Fish's book, which lays out this branch of the thesis. Asking whether literary criticism can address "issues of oppression, racism, terrorism, violence against women and homosexuals, cultural imperialism, and so on," Fish insists that it can do so only if it changes so that it is no longer literary criticism:

> It is not so much that literary critics have nothing to say about these issues, but that so long as they say it *as* literary critics no one but a few of their friends will be listening, and, conversely, if they say it in ways unrelated to the practices of literary criticism, and thereby manage to give it a political effectiveness, they will no longer be literary critics, although they will still be something and we may regard the something they will then be as more valuable. (1)

The circularity is inescapable: literary criticism is that which pertains only to literary critics and their friends. If literary critics acquire any political effectiveness, they can do so only "in ways unrelated to the practices of literary criticism," and therefore will no longer be literary critics, because literary criticism is that which pertains only to literary critics and their friends. The crucial phrase, however, is "unrelated to": how do we know when literary critics are gaining influence in ways unrelated to (or insufficiently related to) the practices of literary criticism? I would have thought, for instance, that my treatment of "delayed" language acquisition in *Life As We Know It* owed something to my training in literary theory (particularly my reading of Wittgenstein), just as my concern with the representation of people with Down syndrome owed something to my reading in cultural criticism. But if my work has any impact on people other than my academic friends, according to Fish, it will be only to the extent that I have written about such things in ways "unrelated" to the practices of literary criticism.

I trust I have managed to make the case that there are uneasy moments and crafty maneuvers in Fish's argument—or, at the very least, so many competing arguments within that argument that it is difficult to know what will count as a "rebuttal" of it. But I do not presume to have made the case that Fish's argument is incoherent. Far from it; the argument is internally consistent at all points, for all of its points depend on an axiom that is central to Fish's work: there is no such thing as "critical thinking." The reason the skills we teach in literary criticism cannot have any impact on the world at large is only tangentially related to the question of whether literary criticism can be unliterary or whether it can travel "regular routes" to state power; the real reason that our skills have no purchase on the world is that there is no possibility of enhancing anyone's capacity for reflection and self-critical analysis.

The idea that the study of textuality might perhaps lead people to envision the world differently, then, is in principle deluded, because the protocols of interpretation (and what Fish calls "reflection in general") do not admit such effects. Hence Fish's especial skepticism with respect to cultural studies, since cultural studies promises precisely to bring skeptical self-scrutiny to bear on academic protocols of interpretation:

> What is true of cultural studies is true of reflection in general, that mode of mental activity of which cultural studies is supposedly the institutional form. It is not that reflection is impossible—most of us engage in it every day; it is just that rather than floating above the practices that are its object and providing a

vantage-point from which those practices can be assessed and reformed, reflection is either (a) an activity *within* a practice and therefore finally not distanced from that practice's normative assumptions or (b) an activity grounded in its own normative assumptions and therefore one whose operations will reveal more about itself than about any practice viewed through its lens. (106)

There's an excluded middle in this formulation, and it's buried so deep that only critical thinking can exhume it: either we're "floating above" our practices (an utter impossibility), or we're resolutely embedded in them, "finally not distanced from that practice's normative assumptions." Like the claim that literary criticism can have extraliterary effects only if it is "unrelated to" literary criticism, Fish's characterization of critical thinking depends on a deliberately fuzzy idea of "distance." Whereas earlier, the question was, "how unrelated do you have to be to be unrelated to literary criticism?" now the question is, "how undistanced do you have to be to be undistanced from your normative practices?"

There are two ways of getting at this problem. The first is to suggest that there are some normative practices, literary criticism and interpretive theory among them, that *entail* scrutiny of one's normative practices. The second is to suggest, less recursively, that the excluded middle is the realm in which we humans actually live whenever we're unsure about what we feel or think. For Fish, you're either in a practice or out of it; and since you cannot scrutinize your practice from outside, you must be doing it from inside, Q.E.D. But perhaps there's a good reason James Joyce, in trying to revive the etymological sense of doubleness in the word "doubt," coined the term "of twosome twiminds" in *Finnegans Wake*. If we are ever to change our minds (or our practices) in any substantial sense whatsoever, there must be some mechanism by which we examine our practices—perhaps even a mechanism by which we try to take the parallax view of seeing the world through two sets of practices, in doubt as to which is the more compelling, of twosome twiminds about what to think or to do.[4] And perhaps, just perhaps, training in the varieties and possibilities of interpretation can enhance one's ability to think of the world through twosome twiminds, to inhabit or at least entertain the possibility of inhabiting more than one practice at a time—and perhaps, if this training travels some irregular routes out of the academy, it might make its way to someplace where it might do some good. Or evil.

This much seems hardly debatable to me (immersed uncritically, as I am, in my Keatsian-Arnoldian practices), but before I move on I have to point out that there's something strangely metaphysical about Fish's polemic against critical thinking, something deeply antipragmatist. I call it "the metaphysics of stasis," and I take it as a canny—but finally unsatisfying—theoretical attempt to distinguish between mere appearance (where it looks as if critical thinking is possible) and bedrock reality (where we can see once and for all that all thinking is an activity *within* and not detached from a practice). It is because Fish is so uncritically embedded in this metaphysics of stasis (indeed, it would seem impossible to be self-critical of it and yet practice it, so Fish's account of thinking, again, has the virtue of being internally consistent) that, as we shall see in chapter 8, his description of "change" is so unconvincing; it is also, incidentally, why his career is distinguished by (among other more notable things) the repeated insistence that he has not substantially changed his mind—as when he directs the reader of *Professional Correctness* not to read the book "as evidence that I have changed my mind or my politics" (x).

I do not want to ask metaphysical questions of Fish's metaphysics of stasis in return; I do not want to reply to his account of thinking, "is it true?" Instead, I want to ask the appropriately pragmatic questions, "does it work?" and "is it useful for us to believe?" And to those questions, there is a wealth of evidence that the proper pragmatist answer is no. The answer is no, I propose, for two reasons—retrospective and prospective. The retrospective reason appeals merely to the historical record, to the fact that there is a vast weight of human testimony to the proposition that transformative reflection *is* possible, that normative assumptions can be changed even by internal normative procedures of scrutiny, that interpretation can modify its object. The prospective reason suggests that if we have any investment whatsoever in the possibility of progressive political change, it would probably be best for us to believe (and therefore ıe" in a pragmatist sense) that human practices can be altered by critical reflection. In this case, then, the mere appearance that people change their minds, that ideas spark revolutions, or that reflection can have material force is sufficient in itself, even if it is "really" only an appearance, to allow us to imagine that there is some potential purpose to the interpretive work we do in the humanities.

What makes my position so reasonable, sane, and ultimately *right,* I think ("right" in a pragmatist sense, of course), is that, unlike many of the academic leftists Fish brilliantly skewers in *Professional Correctness,* I do not maintain that critical work in the humanities *necessarily* enhances the material or intellectual conditions for a more just society. But where Fish catches any number of his colleagues saying preposterous things about how their readings of Shakespeare will be the end of global capitalism, Fish himself produces a mirror image of their political determinism: where they insist that the application of literary-critical skills to extraliterary contexts *will* have political force, Fish insists simply that (to quote the passage again) "those contexts *will* remain unaffected by the application." By contrast, I contend merely that our skills *may* have political force, and that we should proceed (if "we" are progressive-left educators) as if they will. Critical thinking, like the domain of the aesthetic, is "reversible," in Jerry Herron's terms; it can be deployed by William Kristol in the service of a deeper, more searching reading of von Hayek just as it can be deployed by Richard Rorty in a profound reading of John Dewey. As John Guillory writes, with a keen eye on the ambiguousness of the institution (and practice) he inhabits,

> If progressive teachers have a considerable stake in disseminating the kind of knowledge (the study of cultural works as a practice of reading and writing) that is the vehicle for critical thinking, this knowledge is nevertheless only the vehicle for critical thought, not its realization. As cultural capital it is always also the object of appropriation by the dominant classes. (54–55)

Or, in other words, to return yet once more to the issue of employment, if you study English, you will learn how to see through corporate capitalism while qualifying for a job at IBM.

There is one final point on which I find *Professional Correctness* to be an engaging but self-contradictory text, and it has to do with the rhetoric of public justification. On one hand, having studied and written about the history of professionalism, I agree entirely with Fish's claim that "it is a requirement for the respectability of an enterprise that it be, or at least be able to present itself as, *distinctive*" (17). Literary criticism, in other words, must have an identity distinct from history, sociology, cultural studies, anthropology, communications, law, and so forth, *not* because literary criticism is intrinsically unrelated to those fields, but because as a matter of bureaucratic and professional procedure, literary criticism must demarcate some sphere of attention—even under so broad a heading as "textuality"—if it is to survive as an academic discipline. (This was perhaps the signal professional achievement of the generation of critics running from John Crowe Ransom and "Criticism, Inc." to Northrop Frye and *The Anatomy of Criticism:* they had a definite and convincing account of what made literary study distinctive and therefore justifiable as an institutional enterprise.) There are any number of people who believe that the study of literature should not be an academic discipline at all, but, of course, it is not for those people that I write; I write instead for anyone who thinks that literary study might have a purpose for which it could profitably (cough) be "institutionalized" in schools and universities. Institutionalized literary study, as an academic subject and as a profession, simply will not exist very much longer if it does not demarcate, for its potential clients, its domain and procedures, however loosely these might be defined. The alternative is the option I sketched out in chapter 1, namely, the future in which cost-cutting administrators wake up to find that eight different departments are claiming to study "culture," and that therefore six of them can be eliminated and the other two amalgamated.

It is a real surprise, then, that Fish, who is so savvy when it comes to the protocols of professionalism, would close his discussion of "public justification" by declaring, in effect, that literary criticism has no basis— and perhaps no need—for public justification:

> Literary interpretation, like virtue, is its own reward. I do it because I like the way I feel when I'm doing it. I like being

ENGLISH FOR EMPLOYMENT

rought up short by an effect I have experienced but do not yet
\derstand analytically. I like trying to describe in flatly prosaic
rds the achievement of words that are anything but flat and
prosaic. I like savouring the physical "taste" of language at the
same time that I work to lay bare its physics. I like uncovering
the incredibly dense pyrotechnics of a master artificer, not least
because in praising the artifice I can claim a share in it. And
when those pleasures have been (temporarily) exhausted, I like
linking one moment in a poem to others and then to moments
in other works, works by the same author or by his predecessors
or contemporaries or successors. It doesn't finally matter which,
so long as I can *keep going*, reaping the cognitive and tactile
harvest of an activity as self-reflexive as I become when I engage
in it. (110)

I could criticize this passage pragmatically on its face, by noting that it
will probably avail English departments very little to compose official
publicity documents that advertise literary study as a discipline that will
allow Stanley Fish the opportunity to continue feeling good. But I'd
rather ask about the function of "self-reflexivity" in Fish's description of
criticism. If indeed literature is a self-reflexive activity that leads its
devotees, theorists, and explicators to become self-reflexive in turn, might
not this self-reflexivity serve some social purpose with regard to the
composition of civil society?

We will recall from chapter 4 that George Levine has made a strong
case for "the aesthetic" as that which must provide us with our necessary
legitimation "if English, as a profession sustained by publicly and privately endowed institutions, is to survive" (43–44). The aesthetic, in this
sense, is that nebulous thing on which we base our distinctiveness as a
profession, our necessary difference and distance from communications,
history, advertising, and so forth. But in his introduction to *Aesthetics
and Ideology*, which I examined in chapter 1, there is yet another defense
of "the aesthetic" that relies not on the discipline's professional need to
declare the formal uses of language as its domain, but on the possibility,
as I phrased it at the outset of this chapter, that there is a nontrivial
connection between the relative autonomy of "the aesthetic" and the

relative autonomy necessary to the functioning of any critical institution in civil society, including universities. Levine writes,

> [Edward] Said clearly believes—despite his heavily political orientation—in a kind of intellectual free space that might be associated with the realm of the aesthetic. He speaks, at the start of *Culture and Imperialism*, of how, in writing the book, "I have availed myself of the utopian space still provided by the university, which I believe must remain a place where such vital issues are investigated, discussed, reflected on. For it to become a site where social and political issues are actually either imposed or resolved would be to remove the university's function and make it into an adjunct to whatever political party is in power" (xxvi). . . . The aesthetic offers, distinctly, something like the space of the university implied by Said. . . . The aesthetic remains a rare if not unique place for almost free play, a place where the very real connections with the political and the ideological are at least partly short circuited. (15, 16, 17)

This is an altogether different set of claims from the pragmatic Levine/ Fish insistence that "the aesthetic" might legitimate the profession of literary study; in this part of his introduction, Levine argues that the aesthetic is an agent of self-reflexivity both for individuals and for whole societies. It may be, moreover, that there are intimate historical linkages between the category of civil society and the category of the aesthetic as it has been understood since Kant. Levine seems, at least, to be gesturing toward the possibility that the social forces of the eighteenth century, which bequeathed us various forms of nonauthoritarian government and plural public spheres, also created the conditions for a noninstrumental understanding of art as that which serves neither church nor state. Are, then, the autonomies of the aesthetic and of civil society mutually defining and interdependent? This would seem to be the direction in which Levine's argument wants to lead us, but no sooner does Levine speak of "the aesthetic" in terms of relatively autonomous, utopian space than he refigures that space as a large-scale political salve for interpersonal wounds: "As the aesthetic provides a space where the immediate pressures of ethical and political decisions are deferred, so it allows sympathy or, and potential understanding of people, events, things otherwise threatening" (17).

The aesthetic, then, is not a space where sociopolitical concerns are bracketed but a place where sociopolitical hinges are forged:

> I value literature, as whole cultures often do, because — in spite of its endless diversity and refusals to make things comfortable — it is one means to some larger sense of community, to an awareness of the necessity of personal compromise and social accommodation, civilization entailing always its discontents. Part of the value of the aesthetic is in the way it can provide spaces and strategies for exploring the possibility of conciliations between the idiosyncratic and the communal. (19–20)

This would be a remarkably wishful passage in any context, and though I am drawn to its utopian vision of a link between aesthetics, intersubjectivity, and collectivity (that's peace, love, and understanding for you Elvis Costello/Nick Lowe fans), I also know that it has been some decades now since George Steiner and Thomas Pynchon reflected, in their different ways, on the phenomenon of Nazi officers with a fine appreciation of aesthetic excellence. Wishful or not, however, such a passage is simply unfathomable in an essay that questions the legitimacy of drawing broad cultural conclusions from literary texts *and* that asks whether it's worth reading literature if we're reading it for knowledge that "can be discovered through other materials." Why, indeed, if we want to foster harmony, charity, and compromise, should we turn to aesthetic rather than didactic uses of language? And why should we ask the aesthetic to do the work of leading us to some larger sense of community, if indeed we should be asking the aesthetic to do cultural work of any kind?

The simple ugly fact is that if "the aesthetic" is truly (relatively) autonomous from instrumental uses of language, and if the university is truly (relatively) autonomous from state power, then we cannot predict whether the knowledges produced in these precincts will be put to laudatory or regrettable ends. They — the knowledges and their uses — will be constitutionally reversible, capable of contributing to the making of better managers *and* better anticapitalist critics of IBM.

It is possible to read the reversibility of the aesthetic in a more cynical, suspicious fashion. Tony Bennett and Ian Hunter, for example, have turned quite volubly against the aesthetic as a realm of human endeavor, on the grounds that the nebulousness of the aesthetic has historically made it an ideal vehicle for the inculcation of ruling class beliefs. For Bennett and Hunter, in other words, the very indeterminacy of the aesthetic makes it something that can always be redefined to meet the interests of school examiners (or pedagogical authorities elsewhere in civil society), and thus an instrument of mystification—perhaps the very *means* of mystification. As Hunter writes in "Setting Limits to Culture," Romantic aesthetics since Schiller have provided "an *aesthetico-ethical enterprise*," which we should understand in terms of "the loci of an instituted practice aimed at producing a person possessing a certain aesthetico-ethical capacity and standing on which, it is alleged, knowledge depends" (109). For Hunter, accordingly, the project of cultural studies should be "to *restrict* this concept of culture to the specialized practice of aesthetico-ethical self-shaping in which it has pertinence and to begin to chart the limited *degree* of generality it has achieved as a technique of person-formation in the educational apparatus" (115; see also Hunter, *Culture and Government;* Bennett, *Outside Literature*). As a critique of Schillerian-Leavisite aesthetic education, perhaps, this is a plausible (if somewhat sterile and technocratic) description of how aesthetics came to be made into a discipline and a means of discipline; it certainly suffices to explain the pedagogical-professional narrative of *Educating Rita* (1983), in which Michael Caine tells his female working-class student, Rita (who has just handed in her first paper—on Forster's *Howards End,* of all things, since Rita is in some ways an echo of Leonard Bast), "if you're going to learn criticism, you have to begin to discipline that mind of yours." Literary criticism in this film is not just a regime of technical training but also a question of personal perfectability: what Rita enters into is very precisely a discipline of person formation, and it's by no means irrelevant that Rita's course of study eventually earns her fulfilling employment, independence (from a loutish husband), and a better life with some promise of class mobility (and "better songs to sing"). And yet to read "the aesthetic" as if it has *always* been coercive, *s* in the business of making better people, is to misconstrue its ,bility, its elasticity—just as, in emphasizing the F. R. Leavis ,tion of English letters (or the pedagogical scenario of *Educating a*), it is possible to overlook Frank Kermode's comparatively mild and reasonable counterclaim that it is immodest to believe or propose that by training students in literature and criticism we are "improving them, ethically or civilly."

All this reversibility is well and good, you say, and perhaps the aesthetic can be put to liberatory and oppressive ends simultaneously, but—to return to the question with which we started—is it employable? The knowledges, their uses, their users—do they travel in the economy as even Stanley Fish admits they travel (by irregular routes) into the general culture? What, as so many of my undergraduates have asked me over the past decade, can I do with a degree in English?

In chapter 1 I hinted at one answer, when I pointed out that the liberal arts may well be invoked as the best ground for training people who will have to change jobs and careers repeatedly throughout their lives in a post-Fordist economy. As Joseph Urgo, among others, has pointed out, that rationale for the liberal arts is quickly becoming part of the new common sense of American business. Noting that the American Assembly of Collegiate Schools of Business (AACSB) has enunciated new accreditation standards that mandate "fewer credit hours in business and more credit hours in the liberal arts" (135), Urgo writes that the fungibility of a liberal arts education is what makes it so available for this kind of long-term support—or this kind of long-term exploitation:

> AACSB requirements that increase the liberal arts component at business colleges are responding to developments in the business world, to what might be called a postjob environment. As William Bridges argues in *Jobshift: How to Prosper in a Workplace Without Jobs,* the era of lifetime employment is passing rapidly. In a postjob world, workers prepare themselves for a succession of task-oriented term assignments within multiple career paths. The assumption behind the latest AACSB educational requirements is that, while a business education will ensure competency in the present environment, an engagement with the liberal arts

will prepare students for the obscure destinies of a jobless world. (137)

Unfortunately, Urgo is not quite as skeptical of this development as he might be: it takes little imagination, in this "postjob world," to imagine that the liberal arts are being proposed here as the best college training regime for a new workforce that will have to jump from task to task or from layoff to layoff, without the hope of lifetime employment or employment benefits like health insurance and retirement pensions. In this dystopian scenario, the liberal arts become the social glue that accommodates future workers to a post-Fordist economy in which corporate capitalism no longer provides even minimal guarantees of job security or a living wage. This, I have to admit, is not precisely the kind of "reversibility" I have in mind when I speak of possible rhetorics of public justification for literary study.

In the 1993 conference that was the basis for *Higher Education under Fire,* Gregory Jay asked education analyst Michael Apple a long question that, to my mind, got to the heart of one of the definitive ambivalences not only of that conference and book, but also of the academic Left in general: the ambivalence over the idea of English—or education—for employment. Addressing himself to a number of critiques of the purpose of higher education, Jay asked Apple the following question:

> On the one hand, we get a critique that says that what's wrong today is that more and more students are not being given access to higher education, and that this is a class-based and race-based exclusion, which will keep them in poverty and prevent them from attaining the economic mobility and the economic development they have a right to. On the other hand, you provide a critique of vocationalism that argues against seeing the university as a place primarily devoted to training workers. . . . If we're going to negotiate in public with the powers that be, with state legislators, with parents, and with students, we certainly can't do it on the basis of a position . . . that seems to reject out of hand the relationship between work, education, and economic advancement. So how do we negotiate between these two positions of training for citizenship and training for work, without selling ourselves out in one way or another? (165–66)

e's answer starts off winningly—"it's an easy question," he bluffs
5)—but goes on to address the function of the university in the
ntext of commodity exchange that places a premium on specific kinds
of knowledge that can be sold; the class composition of his own students;
and the meaning of what we refer to as "work." It's an interesting reply,
but I'm not reproducing it here because it doesn't, finally, answer the
question. And the reason it doesn't address the question is that Jay had
put his finger on the major structural contradiction that haunts all
academic progressives, but especially those who work in the humanities:
when it comes to thinking about underprivileged student populations,
we want the university to behave as a social force for equal opportunity,
a democratic corrective to the inequalities of capitalism; we want many
things for such students, no doubt, but among the things we want are
good jobs. But when it comes to thinking about training the professional-managerial class (PMC) and the elite, we would rather emphasize
our capacity for helping people deconstruct corporate capitalism than
stress our utility for helping our future-PMC students find some well-
paying jobs at IBM.

I have no ready-to-hand resolution to this dilemma. If Gregory Jay
were to ask me in 1997 the question he asked Michael Apple in 1993 I
would probably have no direct answer either, even after having witnessed,
edited, and reread the exchange many times. But I can make one more
modest proposal, the counterpart to the public legitimation proposal I
made at the end of chapter 1 with regard to the competing imperatives
of "literature" and "culture": among the kinds of employment we should
keep foremost in mind for our students, and among the institutions with
which we should be most concerned as progressive intellectuals, high
schools—our public and private systems of secondary education—seem
to me to be the salient missing term in the discussion as it has been
conducted thus far.

In saying this I do not mean to neglect other careers that have long
been considered appropriate to the skills of English majors. When I
speak to undergraduates on this topic (as I do at every opportunity), I
usually emphasize the fields in which I myself have worked—journalism,
law, advertising, and publishing. One of the reasons I take such exception

to Fish's argument is simply that I believe that literary study can prepare a student for a career anywhere in media—and for a wider variety of media (particularly print media) than majors like communications. But I stress secondary education for institutional reasons. In chapter 3 I foregrounded high school teaching as a career path for graduate students partly because the topic is usually so off limits. Here, by contrast, I want to address secondary education as a career path not for M.A.s and Ph.D.s so much as for our undergraduate students who may take other routes to teacher certification after their B.A. Fully one-quarter of the undergraduates I teach at Illinois are future high school teachers, and I have been teaching undergraduate surveys and writing-intensive literature courses for most of my career. Some of those students are English majors; some major in rhetoric or English education. Almost all are residents of Illinois who plan to go on to careers in Illinois high schools. Teaching those students the skills of interpretation—and offering them a multicultural curriculum that may, *pace* Guillory and Bourdieu, affect the constitution of cultural capital in the K–12 schools—seems to me one of the more important functions I have as a professional critic and as an employee of the State of Illinois. It was no surprise, in fact (though it *was* something of a coincidence), that even as I was completing work on this chapter, my office phone rang and I wound up in a pleasant half-hour conversation with one of my most talented former undergraduates, who was calling me from Chicago to ask whether I would write her a recommendation for her application to teach in Chicago-area high schools.

There are many reasons we theorists in universities undervalue the cultural work of literary study in secondary schools, and I will not dwell on them here; for now, I am less interested in the reasons for the undervaluation than in proposals and practices that can redress this long-standing neglect of our employment relations to the K–12 system. (This is one of the key reasons that I have become, along with my like-minded colleagues, active in the National Council of Teachers of English.) It is true that secondary education is not considered one of the profession's more glamorous constituencies; but then it is also true that there are few constituencies more important to the long-term health of the humanities in the United States. And it is true, as I have stressed throughout this book, that colleges are part of the credentializing apparatus for the professional-managerial class; but it is also true that we are part of a larger educational apparatus to which all social classes have varying levels of access. And, last, it is true, as Andrew Ross has recently said with regard to maintaining or expanding student enrollments in graduate programs in the humanities, that if we are progressive educators we should seek "to seed all kinds of institutions" with "progressive intellectuals" (82); but then it is also true that among the institutions with which we should be most concerned is the vast educational enterprise in which "English" and the "language arts" are still widely—and, both for better and for worse social purposes—gainfully employed. (For that matter, it is not clear why, if Ross is concerned with the production of progressive intellectuals, he should be so narrowly focused on doctoral programs to the exclusion of the rest of the educational apparatus.) There are many places "literary skills" can travel, by regular and irregular routes. Publishing, journalism, law, and media are among those places—and so is the secondary school system which, according to NCTE estimates, currently employs approximately two hundred thousand English teachers.

NOTES

1. In keeping with my traditionalist tendencies, let me cite as a rigorous analysis of interpretation in law, classical theory, and hermeneutics Kathy Eden's recent study, *Hermeneutics and the Rhetorical Tradition*. For close critical readings of manifestos and minstrel shows, see, respectively, Janet Lyon, "Transforming Manifestoes"; and Eric Lott, *Love and Theft*. For an interesting reading of *Lycidas,* see, among other things, Stanley Fish, *Professional Correctness*.

2. This is not gestural. As I write these words, the *New York Times* reports that new federal regulations will prevent legal immigrants with disabilities from becoming naturalized citizens if they are deemed too mentally incompetent to understand the oath of citizenship (see Dugger). This measure took effect, of course, *after* legal immigrants with disabilities were denied federal aid by the Clinton welfare "reform" bill in 1996. In response to that bill, many legal immigrants with disabilities (or their guardians) considered becoming U.S. citizens; admirably, however, the United States will have none of it—at least not for those especially vulnerable people with disabilities (severe

mental retardation, Alzheimer's, and the like) who may not understand all the obligations of citizenship. As Representative E. Clay Shaw (R.-Florida), sponsor of the measure, said, "proposals to restore benefits [to legal immigrants with disabilities] are simply too costly" (A12). Shaw proposes instead a two- to three-year "block grant" program for the states, after which mentally impaired legal immigrants will be cut from the rolls. Fortunately, according to Shaw, this punitive measure will not have long-term negative effects on people with disabilities, because "the death rate will see that that population shrinks in those two to three years" (A12). At last, someone who knows exactly who Scrooge's "surplus population" is, and when—and why—they should die.

3. Elsewhere, Fish admits that there are plenty of regular routes by which intellectuals can have influence on policy; see, for instance, his treatment of libertarian law professor Richard Epstein for a counterexample. Oddly, however, Fish maintains that Epstein has been influential on the Right not "because Epstein's arguments can be mapped directly onto questions of public policy" (this is exactly what I would say, and I will expand on this premise in chapter 10). The reason Fish does not want to tie Epstein's influence to his work's potential for policy making is that doing so would open precisely the door Fish is trying to close: for in this respect what is true of Epstein's arguments "is often true too of literary arguments, especially of the kind new historicists like to make" (52–53). "Rather," Fish writes, Epstein's importance is due to the fact that "in various corners of our society individuals and groups are searching for ways to pursue certain ends and Epstein belongs to a class of people, law professors, to which anyone interested in effecting immediate change is likely to turn" (53). It seems not to occur to Fish, for the moment, that people might turn to academics in order to "pursue certain ends" precisely because the academics' arguments can be mapped directly onto public policy (in this case, right-wing efforts to eviscerate environmental protection laws and employment discrimination laws; Epstein is, for example, cited by Dinesh D'Souza in the course of his proposal to repeal the 1964 Civil Rights Act—for which see chapter 9).

4. For a brilliant discussion of how Fish (and other neo-pragmatists) elide the question of how to account for doubt, see Jules David Law, "Uncertain Grounds."

7

PROFESSIONAL ADVOCATES

WHEN IS "ADVOCACY" PART OF ONE'S VOCATION?

I recently received two student responses to my teaching that shed some interesting light on my classroom practices and my students' expectations. The first was from a student who wrote on one of my evaluation forms that he or she was glad that I had discussed the question of whether gay or lesbian sexuality was an issue in the work of Willa Cather, Hart Crane, and Nella Larsen. The student was pleased that my class even broached the subject, and praised me for being unlike "those politically correct professors who never bring up controversial topics for fear of offending someone." I admit I was not merely happy with but actually amused by this evaluation, since, of course, it is much more common to hear the term "politically correct" hurled at precisely those professors who *do* bring up the subject of gay or lesbian sexuality in the literature classroom. Then again, I thought, it's also quite common to hear the term applied in the culture at large to people who seem dominated by the imperative not to offend—as when "politically correct" is used more or less as a synonym for liberal hypersensitivity to words like "handicapped," "Indian," or "woman." So here, I decided, I inhabited a nice conundrum: in asking my students whether they thought a writer's sexuality does or does not have any influence on their work or on the way we read it, I was certainly politically correct, in the pejorative sense used by the Right, and, better still, I was also politically correct in avoiding politically correct squeamishness about offending my students.

The second student response got back to me only indirectly. One of our graduate students told me that he had assigned my essay "Public Image Limited" to his class in introductory composition, whereupon one of *his* students asked him whether the Michael Bérubé who'd written that essay was the same Michael Bérubé who taught English at the University of Illinois. Upon learning that the two of us were indeed one and the same, the student was mildly astonished; apparently, he or she had taken a class of mine in the recent past, and would never have guessed my political orientation. At first I was entirely pleased with this report, thinking, well, if there's one thing I'm not guilty of, it's advocacy in the classroom. But then I began to wonder whether in fact I was doing such students a disservice by *not* making it clear to them that I have a stake in American cultural politics, and something of a record of weighing in publicly on various issues of concern to my profession. Not that I should wear my politics on my sleeve or announce my various positions on the NEH, NAFTA, and NATO in the hopes of converting my students to my causes; but perhaps students would be better served if I did not pretend to a form of political "objectivity" I cannot profess and do not even believe in.

I should note that I am skeptical of claims to epistemological objectivity not because I believe that everything is political (on the contrary, I believe that many things are *a*political) but because I believe, with Hans-Robert Jauss and Hans-Georg Gadamer, that "interest" is a precondition for knowledge, and that the surest way to trap yourself inside a narrow, parochial, "subjective" view of the world is to believe that you have transcended all merely subjective worldviews. Indeed, the reason hermeneutics demands of us that we theorize our own historical and epistemological positions is that if we *fail* to do so, if we attribute to ourselves the Archimedean point beyond history and mere "interest," we will almost certainly lapse into dogmatism and intransigence. When Gadamer critiqued the "Enlightenment prejudice against prejudice," therefore, he did so not to defend parochialism but precisely to guard against it—as any responsible teacher and scholar should do.

Still, the question remains: even if I eschew claims to "objectivity" on hermeneutic grounds, does that mean I am entitled to say anything at all in the classroom, or even to address any topic I desire? In the past, when I have strayed from the syllabus and addressed contemporary politics directly, I have largely confined myself to mentioning or describing various issues, policies, figures, or statements; the only social activity I have ever directly advocated is voting. (Of course, I have no doubt that assiduously paranoid conservatives could find systemic bias in my courses merely because I address some issues and not others, and because I do not condemn communism with every other breath. But my courses try to tackle serious subjects in a fifteen-week semester, and therefore I have no time for placating the demands of assiduously paranoid conservatives.) But what about recent Republican plans to cut student aid? Is it not within my ambit as a college teacher to inform students of such measures, and to urge them to write their elected representatives so as to make their feelings known on the subject? I cannot consider that an illegitimate form of "advocacy," since federal student aid policies directly and materially affect the classrooms in which I teach, and I am certainly within my rights as a citizen to advise students that they should participate in the political process, especially insofar as their interests *as students* may be at stake.

But here's where things get tricky. As a limit case, let's take the hypothetical example of an astronomy professor who used his or her introductory cosmology course as a vehicle for recruiting students to support the Strategic Defense Initiative or Phil Gramm in '96, on the grounds that Gramm's candidacy and SDI—or, if you like, Clinton and his national service program—were materially relevant to the future of introductory courses in cosmology. It might be possible to argue that that kind of political advocacy is clearly illegitimate, since it violates the boundaries of a discipline whose object is the study of phenomena that predate any human social organization; this is, in rough form, the rationale most people rely on when they distinguish the "objectivity" of the natural sciences from the inevitable "fuzziness" of the human sciences. (I myself would not consider it proper to advise students to write their representatives about proposed cuts in the NEH or the NEA, since public policy concerning those agencies does not materially *and immediately* affect the students in my classroom, however much it may affect teaching and learning in the arts and humanities in the long run.) But then let's consider the position of a teacher whose job it is precisely to make judgments about various forms of human social organization: how indeed can such a person eschew "advocacy" and remain a responsible member of his or her profession?

The difficulty of this quandary was brought home to me once my second child, James, entered the public school system in Illinois, which he did at the age of three. For the purposes of the Individuals with Disabilities Education Act of 1975, I am legally Jamie's advocate; this poses few ethical problems either for me or for my classroom, of course, but it has introduced me to the possibility that if I were a professor of education whose work concerned the disposition of what's currently known as "special education," I could not possibly carry out my professional and pedagogical duties without advocating one form of social organization over another. Not merely because my job depended on it, so to speak, but because I could not responsibly represent current research in my field without simultaneously attending to the ramifications of that research for public policy. Indeed, among the public policies I would be called on to adjudicate is the very question of whether "special education" should exist at all, or whether the policy of "full inclusion" offers superior educational programs and potential for people with disabilities.[1]

Now as it happens, the study of literature, the way I practice it, rarely bumps up against controversies in public policy or political disputes over the reauthorization of acts of Congress. Literature is after all one of the fine arts, and not an explicitly social discipline like, say, anthropology, history, political science, or law; and it is on these grounds that cultural conservatives have criticized teachers like myself, who stress the social ramifications of literary works, for underemphasizing aesthetic considerations at the expense of political considerations. But literature cannot avoid being a *representational* art, which is why the ancients, in their wisdom, spoke not merely of its capacity to delight but of its potential to instruct as well. Literature, more than music and dance, tends to be propositional, and on occasion it even contains specific propositions about the disposition of human social organization. I find it impossible, in ordinary classroom practice, to discuss literature in ways that do not involve worldviews, even when I am trying to make the simplest case about authorial intentionality.

Let me take a specific example—culled from a novel I teach with some regularity. The great black critic Darwin Turner once wrote of Zora Neale Hurston's *Their Eyes Were Watching God* that the scene in which Janie speaks her mind to her second husband, Joe Starks, on his deathbed, is a profoundly disturbing exchange, since, as Turner put it, nothing about Starks's treatment of Janie merits the cruelty with which she treats him on his dying day. Turner's judgment reads as follows:

> Either personal insensitivity or an inability to recognize aesthetic inappropriateness caused Miss Hurston to besmirch *Their Eyes Were Watching God* with one of the crudest scenes which she ever wrote. While Joe Starks is dying, Janie deliberately provokes a quarrel so that, for the first time, she can tell him how he has destroyed her love. During the early years of their twenty-year relationship, Joe Starks jealously sheltered her excessively; during the later years he often abused her because he resented her remaining young and attractive while he aged rapidly. But in a quarrel or two Janie repaid him in good measure by puncturing his vanity before the fellow townsmen whose respect and envy he wished to command. Never was his conduct so cruel as to deserve the vindictive attack which Janie unleashes while he is dying. (108)

Not a single one of my students, male or female, has ever agreed with this assessment; most of them have disagreed even with Turner's characterization of Janie and her "attack," and, far from being sympathetic to Joe, were outraged that Janie had failed to speak her mind in twenty years of marriage. But that's not the point. The point is that you cannot even begin to broach discussion of that scene, regardless of what you think of Turner's critical judgment *and* regardless of Turner's invocation of "*aesthetic* inappropriateness" (my emphasis), without reference to some notion of what constitutes normative behavior between a husband and a wife—any more than you can teach Twain's *Adventures of Huckleberry Finn* without engaging the meaning of various representations of race, or *Measure for Measure* without engaging students' understanding of social phenomena like justice or gender. Nor is it possible, as I have argued elsewhere, to broach a book like James Weldon Johnson's *Autobiography of an Ex-Colored Man* without delivering yourself of propositions concerning unpleasant things like the Atlanta race riot of 1906 and the inconceivable ubiquity of the practice of lynching at the turn of the century.[2]

Indeed, in her now classic essay "Vesuvius at Home: The Power of Emily Dickinson," Adrienne Rich wrote of literary criticism as a form of advocacy, whereby she tried to retrieve and revivify Dickinson's claims on our attention—or, as Rich puts it, "I have come to understand her necessities, could have been witness in her defense" (158). And as Rich is to Dickinson and Alice Walker is to Hurston, so, once, was T. S. Eliot to John Donne and Irving Howe to Henry Roth: the critic seeking to engage with the writers of the past, if s/he is a responsible critic, will at least want to make those writers intelligible to an audience of our contemporaries, to tell us why those writers are important enough to be considered integral to the history of human expression. And in my teaching, this principle holds true regardless of the writer I am trying to ventriloquize, be that writer a black quasi-feminist conservative Republican like Hurston, a gay midwestern visionary like Hart Crane, or a devout Catholic southerner like Flannery O'Connor.

This is a principle that, under ordinary circumstances, would go without saying: of course a responsible teacher is expected to be an "advocate" of various writers and their worldviews, even if only heuristically. But in these troubled times this principle does not, actually, go without saying—which is why most criticism of so-called advocacy in the classroom is so slippery and protean. When I have seen professorial advocacy come under attack in recent years, I have found that critics sometimes define "advocacy" to mean a specific classroom practice or pedagogical theory; sometimes the term refers to individual texts whose mere presence in the classroom is thought to entail unacceptable political ramifications, like *I, Rigoberta Menchú* or *Their Eyes Were Watching God;* and sometimes the term applies to entire disciplines or subfields. I suggested earlier that if I were a professor of "special education," my job might well depend on whether there continues to be such a thing as special education; likewise, professors in programs of women's studies or African American studies are routinely charged with unscholarly advocacy simply insofar as they advocate the existence—and, on bold days, the growth—of their programs, in a way that no professor of economics would be accused of "advocacy" if he or she advocated the continued existence of departments of economics (or, for that matter, even if he or she advocated the continued dominance of so-called classical free-market models of economics in their field).

For a particularly slippery example of how ordinary scholarship can be refigured as "advocacy," let us turn to a 1995 *Wall Street Journal* op-ed by Catholic University history professor Jerry Muller, who cautions conservatives to think, before they defund the NEH and the NEA, about how those much-maligned federal agencies may have actually *slowed* the spread of feminism and multiculturalism—"advocacy" movements fostered by radical organizations like the Ford and Rockefeller Foundations. According to Muller, apparently, it is acceptable for women's studies programs to study women, but not to advance feminist theory (he refers to "*ever more abstruse* varieties of feminist theory," but I submit that there are those who look upon even garden-variety brands of feminist theory as forms of "advocacy," and that Muller may in fact be among them):

PROFESSIONAL ADVOCATES

As those who follow these matters know, the two philanthropies most active in supporting the humanities, the Ford Foundation and the Rockefeller Foundation, have for over a decade funneled their considerable largess into promoting multiculturalism, programs in women's studies (the institutional incentives of which have diverted scholarly attention from the laudable aim of the study of women to the lamentable pursuit of ever more abstruse varieties of feminist theory), and the burgeoning field of lesbian and gay studies. Conservatives must keep these in mind when making policy recommendations regarding the NEH. (A14)

I am not sure, given Muller's terms, what I would do if I were in women's studies and the women I were studying were themselves feminists; I surmise from this formulation that it is all right to advocate the study of women so long as the women in question don't sound like Mary Wollstonecraft, Virginia Woolf, or (Heaven forbid) Adrienne Rich, who are, of course, advocates of varieties of feminist theory—advocates who attract most fire from conservatives, of course, when they're at their *least* abstruse.

And as I've suggested, matters become thornier yet when "advocacy" is an integral part of one's field. In a brief essay published in the *Chronicle of Higher Education,* Stephen Meyer, a professor of political science at MIT and a conservation commissioner in Massachusetts, writes in reference to proposed revisions of the Clean Water Act, "Any study that holds the potential to shift policy, redistribute resources, and influence the relative power of advocates and opponents of environmental protection is *fundamentally political.* . . . For scientists to pretend to be above the political fray is to consign science to irrelevance in policy making" (B2). Since the study in question, released in May 1995 by the National Academy of Sciences, touched on matters at once scientific and political, and since the House Committee on Transportation and Public Works deliberately rushed the bill to a vote in order to beat the release of the study, Meyer charges scientists with acting unethically because, as he writes, "the panel of the academy working on the report had refused to discuss its details ahead of the official release, for fear of appearing 'political' " (B2).

Meyer's point, and mine, is that some forms of advocacy are not merely *permitted* but positively *mandated* by certain fields of study. Interestingly, Meyer blames narrow professionalism for this state of affairs: professionalism substituting for public-mindedness, academic scientists overly concerned with "publishing an article in *Science* or *Nature*, or giving presentations at professional conferences" (B2). But I would say that the problem with academic scientists who pay no attention to the social ramifications of their work is that they're unprofessional. And what I want to suggest by saying this is that we should rethink what we mean by professionalism when we talk about issues of advocacy and professional responsibility.

In the work of cultural critics influenced by Russell Jacoby, "professionalism" is usually a synonym for mere careerism, an attitude of hermetic self-enclosure that leads academics to think in terms of padding their résumés and accumulating perks rather than advancing the public good. I want to suggest, however, that professionals are supposed to serve clients, and that a professional who does not do so is, strictly speaking, unprofessional. In Meyer's example, scientists whose professional domain touches on the disposition of public funds and natural resources have an obligation to serve the public good as they see it—and that obligation to one's potential clients and constituencies should not be presumed to end at the threshold of the classroom.

My argument, then, is that we must recognize that there are innumerable disciplines and subfields in which political "advocacy" for one form of social organization or another is an integral part of one's professional protocols. Conversely, there is another sense in which "advocacy" is simply the name for whatever practice seems to *violate* the professional protocols: at the turn of the previous century, for instance, teachers of evolution were considered practitioners of political advocacy. Therefore, just as there is a sense in which professors of special education or women's studies are compelled to be advocates, so too is there a sense that in astrophysics as practiced in the 1940s and 1950s, advocates of the Big Bang theory were seen as engaged in a form of special pleading that violated the range of reasonable inferences that could be drawn from the available data. As Sidney Hook suggested twenty-five years ago, then, the question of advocacy is always and everywhere a question of professional legitimation:

> The qualified teacher, whose qualifications may be inferred from his *[sic]* acquisition of tenure, has the right honestly to reach, and hold, and proclaim any conclusion in the field of his competence. In other words, academic freedom carries with it the *right to heresy* as well as the right to restate and defend the traditional views. This takes in considerable ground. If a teacher in honest pursuit of an inquiry or argument comes to a conclusion that appears fascist or communist or racist or what-not in the eyes of others, once he has been certified as professionally competent in the eyes of his peers, then those who believe in academic freedom must defend his right to be wrong—if they consider him wrong—whatever their orthodoxy may be. (36)

There's much to admire in Hook's formulation, not least of which is the fact that so few academic or nonacademic conservatives would dare to second it today. What's all the more remarkable about it, however, is that Hook used this rationale to defend a young, impolitic Marxist named Eugene Genovese, who had recently made public his support of the Viet Cong—and, as Hook notes, became immediately infamous for doing so: because New Jersey's Democratic governor rightly refused to fire Genovese from Rutgers on the grounds of aiding and abetting the enemy, the Republican gubernatorial candidate "focused his entire campaign on the issue of Genovese's right to teach" (42). I suggest we will wait in vain for the day when Genovese extends a similar professional courtesy to those "politically correct" scholars with whom he disagrees. Nonetheless, Genovese should have learned an important lesson from this episode, and so should we: our task is not to ask whether "advocacy" constitutes an acceptable classroom practice, of what, for whom, and by whom; rather, our task is to ask each other across the disciplines, from the natural sciences to the human sciences to the professional schools, what kinds of "advocacy" are legitimate—and, in fact, required—by the standards of responsible professional behavior.

In drawing this conclusion, I am not merely calling for academics to have still more conversations about what it means to be academics.

Rather, I am calling for academics to come up with specific and substantive defenses of academic freedom against the incursions of conservative activists who seek to delegitimate entire methodologies and even entire fields on the grounds that they constitute unacceptable forms of "advocacy." Such was the argument of the book that grew out of Lynne Cheney's final pamphlet-salvo from the chair of the National Endowment for the Humanities, *Telling the Truth:* certain "activist" faculty are abusing the principle of academic freedom in such a manner that they can be stopped only by equally "activist" trustees, legislators, and/or alumni groups. (Liberals and principled social conservatives will note that this is precisely the same argument as the claim that certain political dissidents are abusing the principle of free speech and must be censored for the good of the Constitution.) And such is the premise of the latest conservative activist group in academe, not coincidentally led by Cheney and Jerry Z. Martin, the National Alumni Forum. The NAF seeks to bring pressure to bear on liberal and progressive faculty not from within the faculty proper, as is the goal of the National Association of Scholars, but by encouraging alumni and trustees to censure and/or defund "inappropriate" courses and curricula. The guiding idea behind the NAF is simply this: that trustees, alumni, and parents should not support (either financially or politically) the dissemination of knowledges with which they disagree. The NAF is the clearest distillation to date of the ethos of Olin Foundation head and former treasury secretary William Simon, namely, that the folks who pay the piper get to call the tune.

The National Alumni Forum does not portray itself in this way, of course; it promotes itself as bringing "standards" back to academe, and has garnered a great deal of media attention for its "study" that shows how colleges are jettisoning Shakespeare in favor of popular culture. (The survey itself is fraudulent on two counts: one, it compares *required* courses on Shakespeare with *optional* courses on more contemporary or ephemeral subjects, and two, it only counts English department requirements, thus leading its study to the absurd contention that Columbia University, for example, does not have a Shakespeare requirement even though it has two year-long core courses required of all undergraduates as well as a formidable "coverage" requirement of all English majors.)

But the actual appeal of the NAF to wealthy conservative alumni is more forthright: *you should not be paying for these courses on Madonna and gender; you should not have to support illegitimate "advocacy" programs like African American studies.* To many people, in fact, this appeal sounds reasonable enough. Why, after all, should I write checks to a university some of whose faculty criticize private enterprise, when private enterprise has made my generous donation possible in the first place? Why should I not join ranks with like-minded fellow citizens, and try to prune "advocacy" courses from American higher education—or at least from the curriculum of my dear alma mater?

In coming years and decades, I think, progressive educators will not only have to make a principled case for forms of "advocacy" that are intrinsic and necessary to disciplinary formations ranging from political science to biochemistry to special education; we will also have to make a much broader case to the general public, a case to combat the notion that the payer of the piper should be granted his or her every request. That case will doubtless have to look something like this: the distinction between societies that *do* foster knowledges that the wealthy and powerful dislike, and societies that subject such knowledges to the *approval* of the wealthy and powerful, is an absolutely fundamental distinction. It is the distinction, in fine, between free and totalitarian societies. The idea that wealthy alumni and trustees, or elected officials of the state, should be given substantive veto power over the content of a college curriculum is a profoundly authoritarian idea, an idea consonant with autocracy and plutocracy but not with democracy. For those of us college faculty who "advocate" democracy, then, the question of scholarly advocacy will be inescapable—and we should be prepared to advocate for academically free societies at home and abroad.

NOTES

1. Villa, Stainback, Stainback, and Thousand have been among the most energetic of scholarly advocates for "full inclusion" policies, arguing that separate educational facilities are always unequal and therefore always stigmatizing; Carlberg and Kavale, by contrast, advocate "inclusion" in some cases and not others, on the basis of their review of fifty

independent studies of special classrooms, which found that "special classes were . . . significantly inferior to regular class placement for students with below average IQ's, and significantly superior to regular class for behaviorally disordered, emotionally disturbed, and learning disabled children" (Carlberg and Kavale, quoted in Fuchs and Fuchs 526). Of course, the very categories "behaviorally disordered," "emotionally disturbed," and "learning disabled" are themselves open to contestation, such that the constitution of those categories depends radically on our social and professional construction of them. The difference between Stainback et al. and Carlberg and Kavale, in other words, cannot simply be attributed to different "subjective" readings of "objective" data. If I were a professor of special education, therefore, I could not take any stand whatsoever on this set of issues without "advocating" one form of study—and, consequently, one set of findings, and one form of social organization—over another.

2. For my discussion of Johnson, see *Public Access,* 253–62.

8

FREE SPEECH AND DISCIPLINE

THE BOUNDARIES OF THE MULTIVERSITY

Michael Bérubé and Janet Lyon

Both of us have long job titles. We both hold appointments in the English department at Illinois; in addition to this, one of us has a faculty appointment in the women's studies program while the other is a faculty affiliate of the Afro-American Studies and Research Program; and we're both appointed faculty members in the Unit for Criticism and Interpretive Theory. We begin by noting our multiple institutional identities because they bear, directly and indirectly, on the subject of our essay: the ongoing critique of disciplinarity and institutionality that has characterized American universities at least since the early 1960s, and that has issued in the development of such programs as women's studies and African American studies, at the same time that it has made possible the English department careers of transdisciplinarians like Edward Said, Tania Modleski, N. Katherine Hayles, and Homi Bhabha.

In this essay we will discuss two related tendencies in higher education that have become increasingly visible in the three decades since 1964: first, the shift, within the university, from disciplinary structures to interdisciplinary formations, in the shape of cross-listed courses, extradisciplinary units of study, multiply appointed faculty members, and multidisciplinary courses; and, second, the concomitant shift, within English departments, from the study of literature to the study of culture. We're not aiming for any earth-shaking conclusions about these changes in the academy; we're more interested in making some observations about the historical imperatives that lie behind them and the so-called postmodern condition of which they are said to be part. But by our essay's end we hope to have secured, among other things, an understanding of the status of "disciplinarity" in English partly by way of understanding some of the changes in the ways American universities have managed disciplinarity over the past thirty years.[1]

We'll begin by citing a recent theory of institutional change. In an essay entitled, appropriately enough, "Change," Stanley Fish—who was once a literary critic but is now a theorist of law, professionalism, institutionality, interpretive theory, and free speech—addresses the charge that his theory of interpretive communities has no satisfactory account of change. "From the right," he writes, "comes the complaint that an interpretive community, unconstrained by any responsibility to a determinate text, can simply declare a change without consulting anything but its own desires. . . . From the left comes the complaint that an interpretive community, enclosed in the armor of its own totalizing assumptions, is impervious to change and acts only to perpetuate itself and its interests" (*Doing What Comes Naturally* 142). Fish's answer to these charges takes the form of a particularly Fishian reversal and displacement: an interpretive community is not a bulwark against change but in fact an "engine of change" (150); at the same time, it remains a community of shared principles, "homogeneous with respect to some general sense of purpose and purview, and heterogeneous with respect to the variety of practices it can accommodate" (153).

The way Fish redefines interpretive communities as "engines of change" has profound implications for how we understand disciplinarity and the history of disciplinary change: disciplines respond to external challenges, it seems, only if those external challenges are internal to the discipline.

> That is, in order for a formulation from economics or mathematics or anthropology to be seen as related to a problem or project in literary studies, literary studies would themselves have to be understood in such a way that the arguments and conclusions of economics or mathematics or anthropology were already seen by practitioners as at least potentially relevant. To put the matter in what only seems to be a paradox, when a community is provoked to change by something outside it, that something will have already been inside, in the sense that the angle of its notice—the angle from which it is related to the community's project even before it is seen—will determine its shape, not *after* it is perceived, but *as* it is perceived. (147)

Fish's example is drawn from literary criticism and linguistics: from 1957 to 1970, during the period of the so-called Chomskian revolution, "only a small percentage of those working in literary studies was markedly affected by transformational grammar, which came and went without changing at all the way most literary business was done" (147–48). To those who would object that this model is still "too narrowly institutional" and says nothing about challenges from outside the disciplines altogether, Fish replies that extra-academic events can influence academic disciplines if and only if those disciplines have the potential to take stock of those events in the first place:

> it depends on the extent to which the members of the community see the event in question as one that has a direct bearing on their conception of what they do; and that will depend on whether or not their conception of what they do, their sense of the enterprise, is bound up in an essential way with political issues. Some of us changed our teaching methods and our research priorities markedly during the Vietnam War; others of us went on as before as if nothing were happening. (149)

So far this is uncontroversial—and, in fact, a good part of what we say in this essay will consist of detailed elaborations on it. But before we proceed we want to sketch out some of the problems with which Fish's essay leaves us. We'll confine ourselves to three. First, Fish's essay remains uncertain about precisely how the parameters of a community or a discipline respond to or accommodate change, since his definition of "community" proves to be as fluid as the definition of "paradigm" in Kuhn's *Structure of Scientific Revolutions:* at times an interpretive community seems to consist of literary critics interested in linguistics; at times it appears to be made up of all professors of literature; and at times it is made up of all professors, or all intellectuals (with regard to, say, nonintellectual phenomena like the Vietnam War). So the size and constituency of these communities are themselves matters for constant interpretation and negotiation. Second, this account would leave us with disciplines whose changes have always already (potentially) occurred within even though the disciplines are not thereby understood to be static: on one hand, "since an interpretive community is an engine of change, there is no status quo to protect, for its operations are inseparable from the transformation of both its assumptions and interests" (156), but on the other hand, change was always prepared for, and new theories therefore change nothing: "A theoretical pronouncement is always an articulation of a shift that has in large part already occurred; it announces a rationale for practices already in force" (155). As we'll see when we get to Berkeley a few pages hence, this position is crucial for understanding the recent history of higher education. And third, as Fish suggests in the closing pages of his argument, different interpretive communities and different disciplines have different investments in seeing themselves as stable or as fluid; whereas practitioners of literary study like to tell themselves that they have utterly revolutionized the field roughly every three weeks, lawyers and scientists inhabit disciplines that may be every bit as turbid and innovation-driven, but that constitutionally "see continuity where others, with less of a stake in the enterprise, might feel free to see change" (157).

So change does and does not happen all the time, and it all depends on which interpretive community you're talking *about* as well as which community you're talking *from*—assuming that you can tell what's inside and what's outside in the first place. It's not clear what conclusions one can draw from this characterization of intellectual change, particularly for an enterprise like English, which, as Fish rightly notes, is so various and capacious that certain areas of the discipline can seem to be undergoing revolutionary redefinition while other areas, barely contiguous, seem to feel no effects at all. Perhaps the one clear conclusion we *can* draw from Fish's model of change is that it helps clarify the English department's striking lack of disciplinary clarity: English, in this model, is a field whose center is nowhere and whose circumference is everywhere. To say so is to put matters somewhat hyperbolically, of course. But it is precisely in this hyperbolic sense that English has come to embody and dramatize the postmodern crisis of higher education more generally: English is a leading player in the current debates over the social function of the university not only because its internal heterogeneity seems to license the discipline to speak for the humanities—and sometimes even for the university—as a whole, but also because that heterogeneity emblematizes the crisis of representation in which American higher education is currently caught.

In order to make this point in more detail we need to back up a few steps and return, as so many of us are charged with longing to do, to the 1960s.

In his famous 1963 position paper *The Uses of the University,* Berkeley president Clark Kerr formulated in simple terms what he took to be the broad public misperceptions about the institution over which he presided: "There are two great clichés about the university," he wrote. "One pictures it as a radical institution, when in fact it is most conservative in its institutional conduct. The other pictures it as autonomous, a cloister, when the historical fact is that it has always responded . . . to the desires and demands of external groups. . . . The external reality is that [the university] is governed by history" (94–95). This sense of how universities are "governed" involves an especially ambiguous claim, and Kerr was called out on it a year later by militant students at his own university. Kerr's blueprint for what he called the new "multiversity" directed the public as well as the university constituency to adapt to a newly paramount aim of contemporary education, namely, to provide students and disciplinary units with the intellectual training and resources designed to service the needs of government and industry. Outraged students, already alarmed by what they saw as research universities' complicity in Cold War escalation, charged that Kerr's "multiversity" amounted to little more than a "public utility serving the purely technical needs of a society" (Free Speech Movement 213) by producing "enormous numbers of safe, highly skilled, and respectable automatons" (Cleaveland 75) who were not educated so much as indoctrinated through a "random series of isolated training situations" (Davidson 278). At best, they claimed, the multiversity's putative service to "history" took "history" in a perversely narrow sense, to mean "a particular stage . . . of American society" (Savio 218) characterized by the production needs of an industrial, military, and technological boom. At worst, it expunged the broader understanding of history that prevailed among activist students at Berkeley, who perceived that the "history" responsible for the accelerated campus production of industrial-military technology was actually part of the momentous dialectic of the so-called world-in-revolution, ranging from Algeria to Mississippi, from Cuba to Beijing.

The American educational policies that produced the fragmentation and compartmentalization of training did so, claimed its New Left critics, in tandem with the specialization necessary for capitalism's dominion. In fact (it now sounds strange to say), many of these critics rejected the whole concept of multiplicity that underwrote Kerr's description of the multiversity; and in doing so they found themselves walking a rhetorical tightrope between, on the one hand, a then fashionable Sartrian denunciation of the brutal effects of fragmentation on individualism, and on the other hand a growing (and peculiarly American) radical democratic imperative to honor cultural pluralism. Ultimately what these students agitated for was a system that would unify knowledges produced in disparate academic sites, render them choate, and especially connect them to knowledges produced *outside* the university. As one activist wrote, transformation "cannot take place unless it occurs *within* and is organically connected *to* the practice of a mass radical *political* movement. . . . Every attempt should be made to connect campus issues with off-campus questions" (Davidson 282).

This problem of how to articulate university knowledges to external social knowledges became the centerpiece of the Free Speech Movement at Berkeley in 1964, and at the heart of the problem was an inflamed debate over the literal and ideological boundaries of the university: the twenty-six-foot strip of land at the perimeter of the campus, where students had been accustomed to setting up publicity tables for extra-university political organizations such as CORE (Congress of Racial Equality) and SNCC (Student Nonviolent Coordinating Committee), was peremptorily reclaimed by the university as an *on-campus* space, suitable only for the circulation of material pertaining to "internal" concerns. In other words, the *literal boundary* between inside and outside became the site of ideological contestation, and the resulting riots and sit-ins that prevailed for the next four months were aimed at breaking the administrative authorization of what constituted "inside" and "outside"—as well as the (corresponding) administrative definition of what constituted "appropriate" topics for discussion within this contested public space of the university.

From this challenging of the university's sovereign "domain," this testing of arbitrary boundaries between forms and sites of knowledge, it is a short step to proclamations issuing from student manifestos like the French Nanterre Manifesto of 1968, which declares that as a result of the increasingly permeable membrane between the insides and outsides of universities, the "critique of cultural alienation" begun in the university is finally "merg[ing] with the critique of socioeconomic exploitation and political oppression" that lies beyond its walls (270). If this process is to be secured, the manifesto advises, "knowledge must . . . ceaselessly be rescued from falling into the status of a thing known; its compartmentalization (in departments or fields of specialization within the Faculty, for example) must be thrown into question; its ultimate goal must be under constant suspicion" (273).[2] Now divisions *within* the university are suspect; the whole notion of departmental solvency is under question. And from the perspective of this university-generated social critique,

FREE SPEECH AND DISCIPLINE

department faculty members are particularly culpable for the state of knowledge:

> The Faculty is at the center of two grand operations directed at the means of understanding and expressions: their defusing and their recuperation. Their defusing is the Faculty of Dead Letters; their recuperation, the Faculty of Human Relations. In the first case, intelligence and inventiveness are subverted from practice toward fetishism of the finished work, of the past, of what is established; in the second case, these qualities [i.e., intelligence and inventiveness] are employed to condition the work force, to increase its efficiency. Defusing creates erudition, recuperation expertise. (271)

On the one hand, a fetish for dead letters that turns education into a precious antiquarianism; on the other hand, education for employment and nothing more than employment. On the one hand, a university that has too little to do with the history of its own time; on the other hand, a university that plays just the wrong role in the history of its own time. In other words, from this angle, the new utilitarian "multiversities" have it within their fractionalizing power to simply recuperate social critique; and the old-style universities, still impervious to history and social critique, refuse to recognize the legitimacy of dynamic contemporary claims on the university.

If you tone down this manifesto and modify its terms of opprobrium, what you'll wind up with is more or less the outline of Jean-François Lyotard's influential monograph *The Postmodern Condition*—a document not translated into English until 1984. In the Nanterre document's distinction between the Faculty of Dead Letters and the Faculty of Human Relations, that is, we can see the outlines of Lyotard's distinction between the "emancipationist humanism" of the old university (a humanism now degenerate and aestheticized, in the Nanterre model) and the legitimation of education through "performativity" in the postmodern university. What Lyotard learned from his days as a protest organizer at Nanterre, we suggest, was just this distinction between the university of Cardinal Newman and the multiversity of Clark Kerr. When "the desired goal becomes the optimal contribution of higher education to the best performativity of the social system," writes Lyotard, two things happen: first, knowledge is put in the service of enhancing global economic competition; and second,

> higher learning will have to continue to supply the social system with the skills fulfilling society's own needs, which center on maintaining its internal cohesion. . . .
> . . . In a context of delegitimation, universities and the institutions of higher learning are called upon to create skills, and no longer ideals—so many doctors, so many teachers in a given discipline, so many engineers, so many administrators, etc. The transmission of knowledge is no longer designed to train an elite capable of guiding the nation towards its emancipation, but to supply the system with players capable of acceptably fulfilling their roles at the pragmatic posts required by its institutions. (48)

There's no question that Lyotard's diagnosis of "the postmodern condition" is deeply informed by the struggles over the multiversity in the 1960s, but we do not point this out simply to condemn Lyotard as a practitioner of failed '68 philosophy.[3] On the contrary, in what follows we want to ask what we hope is a more productive (and certainly a more self-reflexive) question: what happens when Lyotard's account of knowledge in the postmodern era *itself* becomes influential in the humanities, such that various academic disciplines take up the challenge to "dereify" knowledge and contest the criterion of performativity? What happens when this form of New Left critique is itself institutionalized?

It's possible to answer this question simply by saying that if New Left critique has been engaged to any extent by academic institutions, then to that extent it has been defused, recuperated, incorporated, incarcerated. That conclusion would lead you to the leftist version of antiprofessionalism in which "knowledge," like "culture," is an organic whole that gets carved up by evil disciplines and professions flourishing in the wake of multiversities and seeking to consolidate and fortify their private domains. But surprisingly, this position conflicts in telling ways with Clark Kerr's conception of disciplines: far from construing academic departments as privatized domains of knowledge production, Kerr saw them as equal parties in an ensemble that has made the university "a prime instrument of national purpose" (87). That is, where the antiprofessionalist critique of the post–1968 multiversity mistakenly sees professionalism as the barrier that insulates the university from public life, Kerr (rightly) saw professionalism as the very device that increasingly *integrates* the university into the machinery of government and industry. The question, then, is not whether the university will serve the general public; the question is *which* structural and economic segments of the public will be served—and interpellated—by which academic disciplines.

Certainly the early student manifestos like Nanterre and, in this country, the SDS's Port Huron Statement were alive to this nationalist orchestration of disciplines. Disciplinarity in those works is rendered as an explicitly political effect, and is linked to paternalistic authority: disciplines discipline—and in doing so produce subjects of the national military state. The manifestos' rhetorical tropes cast unruliness as unfettered antidisciplinarity, fraternity as revolutionary antipaternalism, and the reform of the curriculum (literally, "the course of events") as a species of public activism. The concluding sections of the Port Huron Statement, for example, declare that

> Social relevance, the accessibility to knowledge, and internal openness—these together make the university a potential base and agency in a movement of social change. . . . [N]ational efforts at university reform . . . must make fraternal and functional contact with allies in labor, civil rights, and other liberal forces outside the campus. They must import major public issues into the curriculum—research and teaching on problems of war and peace is an outstanding example. They must make debate and controversy, not dull pedantic cant, the common style for educational life. (73, 74)

"Debate and controversy," it is fair to say, *have* characterized the subsequent reform of disciplines, marked as that has been by what Steven Connor describes as the steady flow of "the new theoretical allegiances across disciplines." These allegiances, writes Connor, are accompanied by a breakdown of the links between academic institutions and their national contexts. The language of the development of modern literary criticism began in England with a cultural analysis that sought to rescue and reformulate a myth of national identity to stand against the incursions of anonymous and international mass capitalism. In other areas, especially philosophy and art-criticism, clear and continuous national "traditions" were equally powerful constituents in the rise of academic disciplines. (17–18)

If Connor is right, the institutionalization of academic disciplines was part of an attempt not to atomize, but to *salvage* the "organic whole" of culture whose loss was mourned by the radical students at Berkeley.

We therefore think it's more accurate (as well as more fruitful) to say that the academic incorporation of the New Left critique of disciplinarity has produced considerable involutions in fields, such that some disciplines now take the very institution of disciplinarity, *and* the relations between academic disciplines and their social milieu, as legitimate objects of inquiry. And we offer, as examples of this new mode of disciplinary self-interrogation, the multiple uses to which ethnography has been put in fields such as English, psychology, anthropology, sociology, and media studies. To suggest otherwise, we think—to suggest that the university has simply devoured and digested the New Left's challenges to it in the 1960s without departing in the least from its rationalizing, technocratic agenda—is to ignore the establishment and endurance of extradisciplinary programs (again, women's studies and African American studies are appropriate examples here), programs that, in the best of all academic worlds, play a role in hiring and promotion decisions, and in awarding undergraduate and graduate minors.

Overlooking these changes effectively disables one from producing a cogent account of intellectuals in the contemporary university. Take, for example, a recent book by Carl Boggs, *Intellectuals and the Crisis of Modernity,* which contains within its pages two diametrically opposed accounts of the humanities and social sciences in the age of the multiversity. The first of these, in a chapter entitled "The University, Modernity, and the Diffusion of Technocratic Discourse," sounds as if it had been transcribed from Berkeley and Nanterre word for word, referencing Clark Kerr in order to say that nothing is new under the academic sun:

> Despite a quarter-century of critiques, challenges, and reforms, the multiversity that Kerr had in mind remains the hegemonic form of academic life in the 1990s. And the role of faculty has become even more technocratic than what Kerr described in his book. One could go further: the modern university is the locus of state-corporate management of education that administers and controls the production of knowledge. Virtually all forms of scholarship are saturated with a positivist world-view corresponding to this highly rationalized system. (111)

As Boggs describes it, the modern research university has fostered "a massive proliferation of disciplines and subdisciplines that are fragmented and disconnected from each other" (112), thereby rendering faculty unable to address issues of broad social concern. In such universities, Boggs writes, "the hostility to general, multidisciplinary frames of reference makes it difficult for any particular field of inquiry to establish linkages with other fields or address pressing social and political themes with depth and creative insight: capitalism, bureaucracy, racism, patriarchy, and ecological crisis, to name the most visible" (113). Citing Russell Jacoby as his primary witness, Boggs concludes that "the normal politics of university life invariably asserts itself over all but the most ambitious efforts at intellectual creativity" (117). It would appear, then, that only an extremely ambitious and creative professor of social sciences, like Boggs, could have written such a sentence in an academic book.

The problem isn't merely that Boggs is mistaken about the research university; he's certainly not wrong to note the prevalence of quantification, technical jargon, and statistical analysis in the social sciences that most aspire to being considered "hard" sciences (psychology, sociology, economics), and he's not wrong to suggest that professional activity in those fields is often measured, as it is in the natural sciences, by the amount of external grant dollars one's research has generated. The problem, instead, is that Boggs extends this analysis over every academic discipline, from those most fully integrated into the military-industrial research apparatus to those like women's studies and English where "performative" criteria are harder to come by (and where researchers are rarely granted tenure on the basis of their ability to attract external funding). More than this, however, the problem is that Boggs's failure to ask about the status of his own work in the academic technocracy produces a very strange narrative in which "the university" is the agent of rationalization, but "higher education" is the locus of a new critical intelligentsia whose work *does* contest the reification and instrumentalization of knowledge. Only one chapter after characterizing the university as Clark Kerr's knowledge factory, Boggs writes,

> In higher education, literally thousands of scholars in virtually every discipline have been influenced by themes and approaches drawn from the sixties milieu. The New Left preoccupation with ideology and consciousness, empowerment and community, cultural critique, and the transformation of personal life has entered the discourses of history, sociology, political science, anthropology, urban planning, comparative literature, and art history, in some cases precipitating a crisis of mainstream thinking. . . . This expansion in numbers and influence of critical intellectuals in American life over the past two decades is a development of great importance: oppositional discourse is no longer a rarity, despite more than a decade of Reaganism in national politics. (176)

We think Boggs is right about higher education and wrong about the university (indeed, it's interesting that he associates "dissidence" and "crisis" with the former term and technocracy with the latter). Because we would argue that what happened in the "sixties milieu" Boggs finally acknowledges here is that the humanities and the social sciences took on the challenges of Port Huron, Berkeley, and Nanterre—and they did so because the humanities and the social sciences provide the best disciplinary locations from which to theorize a social context for knowledge and thereby "dereify" knowledge. In this we mean to oppose the human sciences to the natural sciences, in which knowledge has to be reified if normal science is to generate testable criteria for itself. This may suggest that disciplines now distinguish themselves from the natural sciences to the extent they involve hermeneutic accounts of knowledge. That is,

195

historicist or contextualist or feminist or sociological accounts of science are not themselves sciences and do not aspire to the protocols of knowledge in the natural sciences (which is one reason they have met with such hostility from some scientists, as in Norman Levitt and Paul Gross's infamous *Higher Superstition: The Academic Left and Its Quarrels with Science*). It was just such a recognition of the defining role of hermeneutics that guided us when our Unit for Criticism and Interpretive Theory sought to clarify its scope and purpose a few years ago: it was agreed that all disciplines were potentially relevant to the Unit and vice versa, so long as they involve a component of *interpretive* theory—thus excluding purely statistical "theories" in econometrics or quantitative sociology.

It is worth noting that the form of interdisciplinarity characterized by the Unit for Criticism and Interpretive Theory operates through centralization, and in so doing serves an important function in our university. By formally bringing together, through zero-time appointments, faculty members from disciplines engaged in some degree by theorized recursivity, the Unit has helped produce dialogue spoken in a kind of esperanto based in shared hermeneutic practices. This dialogue, occurring not only in weekly seminars but also in cross-listed and jointly taught courses, acts as a hedge against both reification and performativity. Another kind of more-than-disciplinary academic unit—typified by women's studies, African American studies, Latino-Latina studies, and the like—works somewhat differently. These units also bring together like-minded faculty from far-flung disciplines, but their primary goal lies in another direction, that is, to place within a variety of departments a number of jointly hired faculty members contracted to teach a percentage of courses addressing women or gender studies or feminist theory, in the case of women's studies, or African American culture, race theory, diaspora, colonialism, and so forth, in the case of African American studies. And while it might be argued (as critics have done) that individual faculty members in most humanities and social sciences departments would be inclined to teach courses like these whether or not the campus had a women's studies program or an African American studies program, the actual existence of these programs *insures* curricular attention, institutionalizes it, puts money (however meagerly) behind it, and partially diverts curricular control away from individual departments and disciplinary protocols.[4]

Where the Unit for Criticism and Interpretive Theory draws its directives almost entirely from within the academy, women's studies programs and African American studies programs have always maintained ties with outside constituencies. This "external-ness"—precisely the legacy of 1960s inside/outside agitation, as we have noted—periodically becomes the target of criticism by opponents of women's studies programs (for example) who charge that such programs produce mere "pseudoscholarship" in the interest of political activism. To take a recent and egregious example, Daphne Patai and Noretta Koertge's *Professing Feminism* concludes its call for the abolition of women's studies by consigning "advocacy" to the street where it belongs:

> Students and faculty should be encouraged to fight sexism and other injustices on their campuses and beyond, but they should not expect to receive academic credit or tenure for doing so. . . . Advocacy is often appropriate, sometimes necessary, in the street. But in the classroom, the more flexible values of liberal education should prevail. (210)

It is hard to see how the values of liberal education can be called "more flexible" when they explicitly mandate the prohibition of feminist scholarship. Whether some women's studies programs (or some aspects of women's studies programs) are wholly given over to mere "advocacy" we aren't prepared to say, but in the ones we know best, intellectual integrity is actually *reinforced* by an interdisciplinary structure: jointly hired faculty members teach out of their home departments, which are in turn jointly responsible for evaluating scholastic work.

The broader issue at stake here, of course, has to do with the function of critical intellectuals and extra- or interdisciplinary knowledges in the postmodern university. It would be a mistake, we think, to take Lyotard and Boggs at face value, and say simply that the state of knowledge in the postmodern era is defined by the technocratic functions of the multiversity. As Boggs unwittingly demonstrates, this narrative leaves you with no consistent way of accounting for the presence of analyses like

FREE SPEECH AND DISCIPLINE

Boggs's in social science, just as it fails to account for the profusion of critics in English departments who interrogate the institutional status of literary study—Bruce Robbins, Paul Bové, Anne Ruggles Gere, Richard Ohmann, Gerald Graff, and so on. The prevalence of this kind of meta-institutional criticism, we suggest, is the result of how the humanities and social sciences took on the critiques of the university mounted at Berkeley and Columbia and Michigan and Nanterre; and if this is so, then the institutionalization of these critiques has not defused them but *disseminated* them through the disciplines—such that the history of disciplinarity itself is now among some disciplines' objects of study (as in investigations of the "constructedness" of philosophy, anthropology, history, English, law, sociology, and so on).[5]

In our emphasis on the literal "externality" of disciplines we differ not only from Stanley Fish but also, and more markedly, from Antony Easthope. Easthope's *Literary into Cultural Studies* sees the field of English as having been transformed entirely from within by the advent of theory, which, he claims, has rendered untenable literary study's foundational distinction between high and mass culture, "literary" and "ordinary" language. We agree that theories of literariness are not merely coeval with but fundamental to the creation of departments of organized literary study in the Anglo-American world; even in a relatively late text like Northrop Frye's *Anatomy of Criticism* one can see a somewhat strained attempt to sequester literary study from the incursions of other "extraneous" disciplines by insisting on criticism's primary pigment:

> The axioms and postulates of criticism . . . have to grow out of the art it deals with. . . . Critical principles cannot be taken over ready-made from theology, philosophy, politics, science, or any combination of these.
>
> To subordinate criticism to an externally derived critical attitude is to exaggerate the values in literature that can be related to the external source, whatever it is. (6–7)

It is precisely this faith in the organicity of the field—and the capacity of the field to be organized—that the profession of literary study no longer professes. But Easthope is surely mistaken in attributing disciplinary change solely to the rise of theory. If Easthope is right that there has been a paradigm shift in literary study (and though we question the source of that shift we do agree that it has occurred), it surely consists in this: the profession is no longer content to investigate literariness, or that which is intrinsic to the field of literary study, but rather revels in its extradisciplinary promiscuity so long as it is enabled to interrogate literature in its social contexts, investigate literary history as institution, and investigate the history of the institution of literary study. Neoconservatives, we contend, are closer to the mark than is Easthope when they suspect that these developments have their roots in the 1960s. Feminism, Marxism, new historicism, cultural materialism, cultural studies, queer theory: these enterprises share, if they share nothing else, a commitment to theorizing the social context of cultural production—a commitment so broad, in some cases, that literature disappears into the larger field of the social, and is addressed by way of Gramsci and Batman, monarchy and penal codes, race riots and Balinese cockfights.

We realize it could be argued that in turning its attention to culture, English is only doing again what it has always done in the past. We believe there's some element of truth in this—which is why everyone from Terry Eagleton to Steven Mailloux to E. D. Hirsch can claim to have called literary study to return to its roots in the ancient study of rhetoric. For as Eagleton, Connor, Easthope, and many others have pointed out, the premise of English studies was ambiguous from the outset: in Great Britain, with Leavis, Empson, and Richards, as in the United States, with Brooks, Ransom, and Tate, literary study's methodological formalism was always hooked up to a broader theory of the whole, organic culture to be recovered by means of literary study. Surely we will not pretend to believe that Wordsworth or Arnold did not sufficiently interrogate the concept of "culture." But we have no way of knowing whether the emergent dispensation—that is, the dispensation of cultural studies—is definitely either a revolution or a restoration; scholars in the Renaissance can claim that they always studied culture, just as writing instructors and historically minded eighteenth-century scholars can insist they were doing cultural studies before anyone heard of it in the United States; but this may be just one more *trompe l'oeil* of "change," thanks to which it looks to those inside the discipline as if the outside of the discipline really was inside all along.

But what does it matter whether one locates change inside or outside the discipline? Aren't these alternatives illusory, since disciplinary borders are so permeable that their outsides are in and their insides are out? And wasn't the Nanterre Manifesto launched at the humanities and social sciences from the outset, and therefore, as Fish might say, already "inside" the fields it challenged?

We contend, first of all, that it does indeed matter how you tell these stories as tales in or out of school. It matters theoretically, because if there isn't an "outside" to the discipline, then your account of change, like Easthope's or Fish's, can simply adopt Kuhn's structure wholesale, and concoct a narrative wherein paradigm development inevitably produces anomaly and forces revolution regardless of external social variables such as funding, student populations, national agendas, and influential manifestos. And it matters practically, because if you don't have an account of the discipline's "outside," you can't have any sense of its potential constituencies. It's for this reason that we prefer Boggs's account of disciplinary change to Easthope's, because for all its contradictions it rightly foregrounds the social movements of the 1960s, in concert with the expansion of the universities, as precipitates of the contemporary crises in the humanities and social sciences.

Our argument, then, is this. The evolution of the modern multiversity had at its source two principal imperatives: to accommodate new student populations and to produce research (and researchers) of use to national security and global economic competition. Under both headings, the university was put, as Clark Kerr said, into the service of the nation: in democratizing higher education, expanding the franchise to previously excluded populations, as well as in militarizing higher education and providing research and development for the postwar economic boom. Indeed, in the 1990s it is now beginning to look as if the former movement, democratization, was predicated on the latter, militarization. Be that as it may, this vast social change produced the conditions for the dismantling of—or at least the institutional resistance to—the very system of technocratic performativity that sought to manage new student populations as national human resources in the first place. We therefore see the postmodern university as radically porous, accommodating an outside that is its inside, but we do not want to make the mistake of seeing our own disciplinary transformation from literature to culture as a purely intellectual matter of new paradigms and theories developed from within out of the gradual obsolescence of the old.

Our argument does not end here, however. For it would be easy to celebrate the polymorphous perversity of English, and to call an end to the oppressive regime of disciplinarity once and for all, if it weren't for the collapse of the academic job market. If times were flush we could say that English has laid claim once again to being the queen of the disciplines, rightly generating conservative attacks and recruiting new, engaged students to the field in record numbers, and we could be happy, proud, and downright smug. But a bear market as severe as this one throws the profession's standards of value into turmoil, because (as chapter 4 argued) there are now so comparatively few institutional locations for so many different kinds of work. Our ideal aspirations of postdisciplinariness are therefore constantly contradicted by the exigencies of our fiscal austerity. Such a state of affairs, we believe, is harmful not only to the university's intellectual future but also to its present mechanics; that is to say, the contradiction between interdisciplinary expansion and fiscal contraction has grave consequences for graduate programs, hiring and tenure, and peer review at every level—all the forms of professional self-regulation (and intellectual self-definition) we have.

It could not be more important in the current climate, therefore, that universities and legislatures be persuaded to preserve and foster interdisciplinary units—whatever programs, institutes, and centers generate hybrid and critical forms of knowledge whose uses are not foreordained by national policy. We find it bitterly ironic that so post-literary a theorist as Tony Bennett, who has called for greater attention to policy in cultural studies, would criticize progressive intellectuals for opposing governmental calls for "a greater stress on skills and competency trainings in higher education"—as if such a strategy were self-evidently revanchist and reactionary: "the readiness of cultural studies intellectuals to fall into line behind traditional defences of the humanities as a form of education

FREE SPEECH AND DISCIPLINE

which exceeds the mundane calculus that the notions of skills, trainings and competencies imply," according to Bennett, is "most disappointing" ("Being 'In the True' " 235). What's odd about this is that Bennett ostensibly wants to move his field (and ours) outside literature to culture and policy, but in advocating an idea of the intellectual as "technical exemplar" and inveighing against the "useless" knowledges he considers (along with Ian Hunter) to be oppressive regimes of person formation, Bennett seems to have misread the lessons of the past thirty years rather dramatically, so much so that he's now in the anomalous position of proposing that all intellectual enterprises in the humanities be handed over to the proponents of performativity. In opposing the Faculty of Dead Letters, it seems, Bennett has merely proposed that we join the Faculty of Human Relations. No strategy could be more self-defeating for American critical intellectuals in the academy, who need, in these straitened times, every political and rhetorical opportunity to keep open and thriving those interdisciplinary units through which knowledge can circulate such that its directions and goals can be neither predicted nor controlled. And we insist on this commitment to interdisciplinarity not because it is an end in itself, but because we know that in the modern research university, it is one important way to keep alive the attempt to dereify knowledge; it is one of the few ways we can value even those knowledges that do not immediately serve the purposes of the state.

NOTES

1. For a synoptic (and thoroughly interdisciplinary) overview of the history of disciplinarity in higher education, see Messer-Davidow, Shumway, and Sylvan, *Knowledges: Historical and Critical Studies in Disciplinarity*.
2. For a recent and eloquent restatement of the intellectual imperative to dereify knowledge as it is produced in universities, see Readings.
3. See Starr, however, for a thorough account of the role played by the Nanterre and Mai '68 movements with regard to subsequent developments in French social theory.
4. As Ellen Rooney has written, "The creation of women's studies programs entails a specifically feminist critique of the disciplines. This critique is predominantly anti-essentialist and attacks the common-

sense view of disciplinary discourse as at least potentially objective in its representation of the real" (21).

5. This is not to say that the dissemination of New Left critiques in the contemporary university cannot be made to work for the forces of technocracy; as Steven Connor points out,

> The forms of interdisciplinarity which result from this exchange of [critical] languages and [intellectual] concepts are often claimed as postmodern destabilizations of the structures of knowledge. But this argument could be put the other way around. The form of interdisciplinarity which has been fostered across the social sciences and humanities by the vehicle of the postmodernism debate can also be seen as attempts to master the field, coercing it into intellectual performativity. (42–43)

Bill Readings's analysis of "the University of Excellence," in which all disciplines and faculty specializations are welcome so long as they are "excellent" (that is, amenable to quantitative administrative review of some kind or usable for the machinery of "prestige"), suggests that Connor's dystopian reading of interdisciplinarity will be altogether germane to the university of the twenty-first century. At the same time, Berkeley Free Speech Movement activist Bradford Cleaveland's declaration that "the most salient characteristic of the 'multiversity' is massive production of specialized excellence" (75) suggests that Readings's dystopian reading of the university, like Lyotard's, has its roots in Berkeley and Nanterre.

9

EXTREME PREJUDICE

THE COARSENING OF
AMERICAN CONSERVATISM

S trolling through the Detroit International Airport in October 1995 on my way to my parents' home in Virginia Beach, I came upon a newsstand-bookstore that was devoting eight or ten shelves of space — roughly one-quarter, I believe, of its "new best-sellers" wall — to Dinesh D'Souza's *The End of Racism*. I had heard a great deal about the book before it was published, and had just recently been asked (twice, actually) by the *Chicago Tribune* to review the thing. I declined, partly on the grounds that I've already read more D'Souza than any human should, having perused both *Illiberal Education* (1991) and his rarely mentioned first (and best) effort, *Falwell: Before the Millennium* (1984). That's the book where D'Souza writes, "listening to Falwell speak, one gets a sense that something is right about America, after all" (205). So why would I want to read the new seven-hundred-page D'Souza, the *magnum opus,* the D'Souza *Ulysses?* Do I really have any obligation to keep plowing through the bookshelves of the Right, demonstrating again, again, and yet again that there's no there there?

Within hours I was in my parents' living room, asking my father whether he thought a *Tribune* review from me would make any dent in the media campaign bringing bulk shipments of *The End of Racism* to airport bookstores, or whether I wouldn't just be giving the book greater visibility and credibility simply by agreeing to treat it as a serious object of some kind. "Well, Michael," my father replied, "you may not have to worry. From what I hear, the book isn't doing very well, in reviews or in sales." When I asked where my father had heard such a thing, he turned to me and asked, with a straight face, if I hadn't seen the new "desperation" ads the Free Press was running for the book. "Two for one deal," he said. "Buy *The End of Racism* at the already low, low bargain price, and receive *The Mark Fuhrman Tapes* for free."

Of course, it's manifestly unfair to compare D'Souza and Fuhrman. To my knowledge, D'Souza has never personally beaten or framed a black person, nor has he suggested creating a large bonfire of black bodies. In *The End of Racism,* he merely proposes a theory of "rational discrimination" based on the recognition that there are vast "civilizational differences" between black and white Americans. At the close of his first chapter, D'Souza offers a brief catechism on the subject: the main problem for blacks is not racism but "liberal antiracism" (24); the civil rights movement failed because "equal rights for blacks could not and did not produce equality of results" (23); and, consequently, the cause of "rational discrimination" is "black cultural pathology" (24). D'Souza's middle chapter ("Is America a Racist Society?") expands on the premises of rational discrimination, which may be unfair to individuals but valid about groups-as-wholes:

> Only because group traits have an empirical basis in shared experience can we invoke them without fear of contradiction. Think of how people would react if someone said that "Koreans are lazy" or that "Hispanics are constantly trying to find ways to make money." Despite the prevalence of anti-Semitism, Jews are

rarely accused of stupidity. Blacks are never accused of being tight with a dollar, or of conspiring to take over the world. By reversing stereotypes we can see how their persistence relies, not simply on the assumptions of the viewer, but also on the characteristics of the group being described. (273)

This, perhaps, is right-wing sociology's finest moment: *reversal of stereotypes!* why didn't *we* think of that? OK, now let's get this straight. Koreans are *not* lazy, Hispanics do *not* try to make money, blacks are spendthrifts, and . . . hey! wait a minute! those clever Jews really *are* trying to take over the world! Get me Pat Robertson!

Many of my black friends were understandably alarmed to hear that D'Souza's book endorses the practice of "rational discrimination." One told me that she'd read only so much of the book—up to the point at which, on page 169, D'Souza notes that the civil rights movement failed because it did not consider its political consequences, namely, that "racism might be fortified if blacks were unable to exercise their rights effectively and responsibly." After that, she decided the book might as well be called *The Negro a Beast,* after Charles Carroll's best-seller of 1900. Such a title, I replied, would almost surely keep the book out of major airport bookstores, and so was probably rejected by the Free Press's marketing department. But then again, I added, there's no reason to think of D'Souza as antiblack; on the contrary, the theory of "rational discrimination" may prove even more dangerous to white Americans than to any other group. It doesn't take a Malcolm or an Ishmael Reed to figure this one out: White people blow up federal buildings. White people pillage savings and loans. White people built Love Canal. White people commit horrid, unthinkable murders of helpless children and pregnant women, and then they blame them on black men. All the great serial killers of the West are white people. Now, don't get me wrong. I'm not saying that all white people are crazy or greedy or dishonest. Some of my closest friends are white. But would you want your daughter to marry one?

I presume that many of my readers are familiar with some of D'Souza's more extraordinary arguments in *The End of Racism.* Still, it may be worth pausing briefly over some of the highlights. For I believe that this book, together with *The Bell Curve,* is an instance of a wholly new genre of encyclopedic pseudoscience, and it is fundamental to the workings of this genre that the books in question be too bloated and overstuffed for the ordinary reader to fathom (*The End of Racism* features no fewer than 2,198 footnotes, which makes it very scholarly). In this new genre, measured commentary, reportage, and scholarship are blended with ultraconservative and even fascist policy recommendations, regardless of the logical relation between the scholarship and the recommendations. (D'Souza's book differs from Herrnstein and Murray's in that it also includes extended hallucinations masquerading as "historical overview." More on this below.) The authors of these books then appear, calm and composed, on national media, saying they know their work is bound to cause controversy but should at least be granted an honest hearing. (See also, under this heading, David Brock's book *The Real Anita Hill*.) Phase two of their mission accomplished, they then head back to base camp at *Commentary* magazine to write assessments of their reception, showing that despite their honesty and all-around reasonableness, they were savaged and brutalized by the knee-jerk liberal press. All of which demonstrates *a fortiori* the liberal stranglehold on political discourse in the United States; for as the ever reliable Eugene Genovese memorably put it in a 1995 issue of the *National Review*, surveying the public response to *The Bell Curve*, "once again academia and the mass media are straining every muscle to suppress debate" (44).

So much for the new genre and its characteristic media-saturation strategy. Now for some of the highlights of *The End of Racism*.

- "The popular conception seems to be that American slavery as an institution involved white slaveowners and black slaves. Consequently, it is easy to view slavery as a racist institution. But this image is complicated when we discover that most whites did not own slaves, even in the South; that not all blacks were slaves; that several thousand free blacks and American Indians owned black slaves. An examination of these frequently obscured aspects of American slavery calls into question the facile equation of racism and slavery" (74–75).
- "The American slave *was* treated like property, which is to say, pretty well" (147).
- "Most African American scholars simply refuse to acknowledge the pa-

thology of violence in the black underclass, apparently convinced that black criminals as well as their targets are both victims: the real culprit is societal racism. Activists recommend federal jobs programs and recruitment into the private sector. Yet it seems unrealistic, bordering on the surreal, to imagine underclass blacks with their gold chains, limping walk, obscene language, and arsenal of weapons doing nine-to-five jobs at Procter and Gamble or the State Department" (504).

- "Increasingly it appears that it is liberal antiracism that is based on ignorance and fear: ignorance of the true nature of racism, and fear that the racist point of view better explains the world than its liberal counterpart" (538).

Almost as striking are D'Souza's incisive rhetorical questions:

- "If America as a nation owes blacks as a group reparations for slavery, what do blacks as a group owe America for the abolition of slavery?" (100).
- "How did [Martin Luther] King succeed, almost single-handedly, in winning support for his agenda? Why was his Southern opposition virtually silent in making counterarguments?" (196).
- "Historically whites have used racism to serve powerful entrenched interests, but what interests does racism serve now? Most whites have no economic stake in the ghetto" (554).

Yet these are merely the book's most noticeable features—the passages that make a reviewer suppose that the easiest way to slander D'Souza is to quote him directly. *The End of Racism* is not, however, the sum of its pull quotes. More important are its characteristic tics and tropes, which are harder to convey but crucial for an understanding of how the text operates. There is, for instance, the repeated insistence that behind every civil rights initiative looms the specter of cultural relativism, and that the father of cultural relativism is Franz Boas. The last time I encountered this argument—and I am not making this up—I was reading neo-Nazi pamphlets on the cultural inferiority of the darker peoples. D'Souza is unique, however, in finding the determinative influence of the evil Boas everywhere he looks, from the founding of the NAACP to the unanimous majority in *Brown v. Board of Education*.

Indeed, the only figure who comes in for as much abuse as Boas is

W. E. B. Du Bois, apparently because Du Bois was so simplistic as to blame white people for lynchings, Jim Crow, and the race riots of 1906 (Atlanta) and 1908 (Springfield). (Actually, to be fair to D'Souza, his book nowhere mentions those riots.) As D'Souza explains at some length, Du Bois was a cultural relativist. And once you've been tarred by D'Souza as a cultural relativist, there is no hope for you. Everything you say testifies only to your moral turpitude. Henry Louis Gates suggests that it's racist to say to him, "Skip, sing me one of those old Negro spirituals" or "You people sure can dance," and D'Souza replies, "Why are [these statements] viewed as racist? Because contemporary liberalism is constructed on the scaffolding of cultural relativism, which posits that all groups are inherently equal" (268). A century earlier, Du Bois had called for "anti-lynching legislation" and "enfranchisement of the Negro in the South"; D'Souza remarks that "this represented a program strongly influenced by Franz Boas and Boasian assumptions" (190). Johnnetta Cole writes that the "problem" with single-parent households "is that they are deprived of decent food, shelter, medical care, and education," and D'Souza writes that "Johnnetta Cole finds nothing wrong with single-parent families" (519)—and that, more broadly, "leading African American intellectuals abstain from criticizing and go so far as to revel in what they describe as another alternative lifestyle" (518). Houston Baker writes a book claiming that 2 Live Crew was rightly banned in Broward County for obscenity, and sure enough, D'Souza cites him (and his book) as one of the Crew's leading defenders. How can this be? You guessed it— cultural relativism. "Instead of seeking to counter the cultural influence of rap, leading African American figures unabashedly condone and celebrate rap music as the embodiment of black authenticity" (513).

In *The End of Racism,* we find that successful black people are especially whiny (unless they're conservatives, who, "unknown and unrecognized," are "striving heroically to make the underclass . . . worthy of respect" [521]); accordingly, they draw from D'Souza a scorn that is indistinguishable from hatred. In 1993, Senator Carol Moseley-Braun argued that the Senate should not recognize the Confederate flag as the official symbol of the United Daughters of the Confederacy; D'Souza calls her protest "histrionic" (286)—and, because he knows who pays his bills, fails to mention that Jesse Helms made a project of harassing and taunting Moseley-Braun thereafter. In his penultimate chapter he takes up the narratives of middle-class blacks who deal with countless racist slurs and slights every day, and reacts with disbelief to their professions of resentment:

> These are the observations of relatively well-placed men and women: an executive, a government worker, and a college professor. Since no reasons are given that would justify such reactions [i.e., D'Souza did not cite them], one might conclude that we are dealing with cases of people who live in a world of make-believe, in mental prisons of their own construction. For them, antiracist militancy is carried to the point of virtual mental instability. It is hard to imagine whites feeling secure working with such persons. (491–92)

D'Souza's ability to empathize with beleaguered white persons is admirable, and no doubt if he continues to succor the hurt feelings of his powerful white colleagues who don't see why Skip and Lani get so huffy when they're asked to sing "Roll, Jordan, Roll," his career as a prominent right-wing intellectual—and his fellowship from the American Enterprise Institute—is pretty much guaranteed. It was not long after the book was published, in fact, that the *Wall Street Journal* devoted half a page of op-ed space to an excerpt from D'Souza's concluding chapter—the part where he finally gets around to delivering his payload, that is, arguing for the repeal of the 1964 Civil Rights Act.

D'Souza's rationale for repeal is clear: "America will never liberate itself from the shackles of the past until the government gets out of the race business" (545). Now that racist discrimination against African Americans is largely a thing of the past—as D'Souza points out, "all the evidence shows that young people today are strongly committed to the principle of equality of rights" (552)—government action can only produce a justifiable white backlash. Drawing his inspiration from legal scholar Richard Epstein, D'Souza does not worry about freeing the private sector from antidiscrimination laws; for in a truly free market, racial discrimination would not exist at all, since "discrimination is only catastrophic when virtually everyone colludes to enforce it" (539). D'Souza's case in point is major league baseball, about which he poses a truly novel thought-experiment: "Consider what would happen," he writes, "if every baseball team in America refused to hire blacks." Lest we are unable to imagine (or remember) such a state of affairs, D'Souza guides us step by step:

> Blacks would suffer most, because they would be denied the opportunity to play professional baseball. And fans would suffer, because the quality of games would be diminished. But what if only a few teams—say the New York Yankees and the Los Angeles Dodgers—refused to hire blacks? African Americans as a group would suffer hardly at all, because the best black players would offer their services to other teams. The Yankees and the Dodgers would suffer a great deal, because they would be deprived of the chance to hire talented black players. Eventually competitive pressure would force the Yankees and Dodgers either to hire blacks, or to suffer losses in games and revenue. (539)

There's something disingenuous about D'Souza's plans for integration, since D'Souza had argued earlier, citing Joel Williamson, that Jim Crow laws were "designed to preserve and encourage" black self-esteem (179). But let's assume, for the nonce, that D'Souza is serious here, and let's assume also that franchises like the Celtics or the Red Sox of the 1980s could not win games without a sizable contingent of black ballplayers. How precisely is this argument supposed to work in American society at large? Are we supposed to believe that bankers and realtors don't discriminate against black clients for fear that their rivals down the street will snap up all those hard-hitting, base-stealing young Negroes? Or is it that when black motorists are tired of being pulled over in California they will simply take their business to the more hospitable clime of Arizona?

Few commentators have noted that Dinesh D'Souza is himself the most visible contradiction of the Right's major premise in the culture wars, namely, that campus conservatives are persecuted by liberal faculty and intimidated into silence. For here, after all, is perhaps the most vocal

Young Conservative of them all, a founder and editor in chief of the *Dartmouth Review* who's since gone on to Princeton University, the Reagan administration, and lucrative fellowships from the Olin Foundation and the American Enterprise Institute. He is, in short, a phenomenon. No matter how diligently his critics pore through his work, demonstrating time and again that the stuff doesn't meet a single known standard for intellectual probity, *he is taken seriously*. Liberal heavyweight champ Richard Rorty is tapped to take his book apart in the *New York Times Book Review;* Harvard's Stephan Thernstrom weighs in with a trenchant critique in the *Times Literary Supplement.* On the other side of the aisle, both Genoveses stand up to testify to the book's importance, calling it "impressive" and "courageous." D'Souza is denounced and celebrated, defended and reviled. He appears excerpted in *Forbes,* the *Atlantic,* the *American Scholar.* Meanwhile, over on page A4 of the hometown paper there's a story about how the Philadelphia police have terrorized the city's black citizens for years; on page B10, an NFL star's cousin, a young black businessman, has been stopped by highway police and beaten to death. No probable cause, no previous record. No one notices.

Not long ago Michael Lind wrote about what he called "the intellectual death of conservatism," recounting how he watched in amazement as Heritage Foundation founder Paul Weyrich suggested lacing illegal drugs with rat poison—and no one in the room demurred. The publication of *The End of Racism* seems to me a larger version of the same phenomenon: not only a deliberate and at times terrifying attempt to move the center of political gravity as far right as possible, but also so egregious an affront to human decency as to set a new and sorry standard for "intellectual" debate. It is remarkable, I think, that this latest and most virulent brand of postwar American conservatism has so far produced only one defector, only one conscientious objector—the aforementioned Michael Lind. And it is similarly remarkable that D'Souza's book has provoked only one resignation from the AEI—that of prominent black conservative Glenn Loury.

Still, however much I might lament the resolute ideological conformity on the Right, it strikes me as a gesture of political impotence for commentators on the Left to criticize *The End of Racism* for failing to meet any reasonable standard for sound scholarship, informational accuracy, or logical coherence. It's rather like complaining, after your arms have been removed from their sockets, that your opponent has failed to abide by Robert's Rules of Order. Does anyone seriously expect that Lynne Cheney, say, will tender her resignation to the American Enterprise Institute as well, on the grounds that D'Souza has flouted the intellectual standards of which she claims to be the defender? And what of Adam Bellow, son of Saul, who, according to D'Souza, "worked closely with me throughout the preparation of the manuscript" (xi)? Wasn't there anything he could have done to make *The End of Racism* a saner, a more respectable book? Or was he too busy searching the world over for the Tolstoy of the Zulus?

I think it is important that the American Right is now so supremely self-confident, so assured of its control over the direction of public policy and political debate, that no one at the Free Press or the AEI worried whether *The End of Racism* might damage the credibility of conservatism. Such self-confidence is altogether impressive, even sublime. What does it betoken? The *Wall Street Journal* excerpt of the book should probably be our guide. It's significant that the *WSJ* trumpeted only D'Souza's call to repeal the Civil Rights Act of 1964; apparently, the time is not yet right for the *Journal* to reprint neo-Nazi pamphlet material on the omnipresent cultural influence of Franz Boas. But outright repeal of the Civil Rights Act is still unthinkable in American politics; the most the American Right can do, for the moment, is to shoot holes in the Voting Rights Act of 1965, and torpedo those progressive intellectuals, like Lani Guinier, who actually take the legislation seriously. The "race issue" of 1996 wasn't supposed to be civil rights *in toto;* it was supposed to be the abolition of affirmative action, spearheaded by the so-called Civil Rights Initiative in California. Isn't D'Souza jumping the gun? Isn't the Civil Rights Act too ambitious a target?

But perhaps the jumping of the gun is precisely the point. D'Souza's not writing for 1996, or even for 2000; he's writing for generations yet to come. Like Pat Buchanan's 1992 and 1996 presidential candidacies, *The End of Racism* may be a short-term novelty but a long-term success in pushing the rightward edge of the envelope for what can be plausibly considered a substantive contribution to public debate. It is a disgraceful book by any measure, but it may yet be a landmark—even though, like the *Lyrical Ballads* and the paintings of the Fauves, it be maligned and ill-understood upon its first appearance. And who knows? Maybe the times they are a-changing, and soon it will be as common as rain to hear Buchanan-esque presidential candidates allude slyly to the machinations of international Jewry and to see policy analysts guffawing about how ridiculous it would be to create jobs programs for gold-chained, limping black men. Once upon a time Barry Goldwater was considered an extremist—so much so that the presidential race of 1964 was the only election since 1852 in which a Democratic nominee other than FDR won more than 50 percent of the popular vote. Now, with his defense of gay military personnel and his dismissal of personal attacks on Bill Clinton, Goldwater has almost become the custodian of the party's "liberal" wing. What if *The End of Racism,* like Goldwater's nomination, is merely a shot across a bow? What if, by the year 2016, the American Right has carried out Rush Limbaugh's jocular suggestion that a maximum of two liberals be kept alive on each college campus—and those few thousand of us who remain amidst the rubble are sighing nostalgically for the days when there were still liberal Republicans like Dan Quayle who were at least ambivalent about sterilizing populations with measurable "civilizational differences" from whites?

Allow me my phantasmic scenarios. I now live in a nation where any number of mainstream, nationally syndicated columnists can promote Pat Buchanan's presidential candidacy, and surely it will not be long before we see the *Atlantic* cover that proclaims, "Pat Buchanan was right." I live in a nation where it is not considered "extreme" to eliminate capital gains taxes or to turn social programs over to the states so that Republican governors can undo the deleterious effects of the Fourteenth Amendment. I live in a nation where Dinesh D'Souza is lauded as a "courageous, insightful, and eloquent critic of the American social scene" (Linda Chavez) and a book like *The End of Racism* appears on airport bookstore shelves festooned with no fewer than eight testimonial blurbs—including those of Chavez, Eugene Genovese, Charles Krauthammer, and a few token liberals like Andrew Hacker and Gerald Early, who really ought to have known better.

What, finally, does the publication of *The End of Racism* say about the relations between the "responsible" Right and the "extreme" Right? In the wake of the Oklahoma City bombing in 1995, American conservatives were outraged that anyone could have drawn a connection between Rush Limbaugh's or Gordon Liddy's mirthful, hypothetical incitements to murder, and the deadly explosives used by right-wing fanatics. Many on the American Right, to their credit, denounced the bombing—usually a few hours after denouncing those few pinkos and bleeding-hearts who had had the gall to suggest that the bombers might not have been Islamic fundamentalists. Not a single white conservative, however, has voiced any reservations or regrets about the publication of *The End of Racism*. Adam Bellow has not stepped forward to admit that mistakes were made; Bob Dole has not charged that the book will erode our moral fiber; Gertrude Himmelfarb and Lynne Cheney have not confided to *Commentary* their worries that the book may not meet the ideal of scholarly objectivity. Perhaps it will not be considered outrageous, then, for progressives like me to draw the obvious conclusions—that there are no rightward boundaries for what conservatives will consider acceptable public discourse on race, and that the *Wall Street Journal*'s editors are willing to flirt with anything, even cryptofascism, so long as it promises to unwrite federal commitments to social justice. As I contemplate *The End of Racism*, I await the requisite soul-searching on the Right. But in all honesty, I'm not holding my breath.

10

CULTURAL CRITICISM AND THE
POLITICS OF SELLING OUT

My first attempt to write this essay dates from the spring of 1995, and one of the more curious features of its original composition was that it turned out to be anything but the essay I had intended to write. When, in the summer of 1994, I was asked by the English Department at Kansas State University to address the subject of cultural studies and the public sphere, I assumed that I was being asked to do so partly because my work has addressed the relations between academic and popular knowledges: the university in the public sphere, the university as a public sphere. I haven't forged any bold, fresh models of cultural studies, I haven't proposed brand new roles for Western intellectuals, and I honestly couldn't come up with a supple new theoretical account of subjectivity if I tried. But I have written a few essays for nonacademic publications, and these have tried to address some conception of "the public" both thematically and materially. So at first, I thought I might discuss what kind of "selling out" such writing entails, partly because it involves a great deal of negotiating with editors about almost every aspect of an essay, multiple rewrites, and hour-long conversations about individual sentences and paragraphs. The extent (and length) of these negotiations is quite considerable, and I am somewhat surprised that so few people have remarked on the difference between academic and freelance writing in this regard. And in the course of discussing "selling out" in this sense, I thought I would try to put some necessary distance between myself and the new discourse of the so-called public intellectual, by explaining that I do not share that discourse's assumptions about critical language, professionalism, or the history of the American intelligentsia in the twentieth century, or, for that matter, its assumptions about the constitution of "the public."

In other words, I had intended to play on the ambivalence of the phrase "selling out," which could mean either abandoning one's principles and caving in to the demands of the market, *or*, more happily, creating the conditions under which academic cultural criticism could reach so wide an audience as to create what the sports and entertainment industries call a capacity crowd. In that sense, I live to see the day when mass-market bookstores find it impossible to keep adequate supplies of the latest book from Stuart Hall or Michael Denning: just as *The Bell Curve* or *See, I Told You So* has sold out in many stores, so too should intellectuals in cultural studies hope to sell out. Or so I was going to say. My essay, then, as I first envisioned it, would have emphasized the differences between the New Right and the academic Left with regard to the *dissemination and distribution* of cultural criticism: no sooner does Simon and Schuster publish Christina Hoff Sommers's *Who Stole Feminism?* than right-wing flacks like Mona Charen and Harry Stein are singing its praises in syndicated newspaper columns and the "On Values" page of *TV Guide* (a publication whose circulation exceeds even that of *Critical Inquiry,* so I'm told). Of course, I know that the Right has better distribution networks than we do for many reasons, partly because they have all the money and almost full control of the liberal media. Nonetheless, I wanted to say, the academic Left is producing a great deal of quite valuable and searching cultural criticism, but, as Edward Said ("Opponents, Audiences") famously remarked, it doesn't tend to circulate to more than three thousand people, most of whom don't need to be convinced of the merits of the cultural work of the cultural Left.

Now, I will argue something like this in the course of this essay, but over the past few years my relation to the discourse of the public intellectual has changed enough to make it difficult for me simply to propose that cultural studies intellectuals do more "public" writing. As of the fall of 1994 I had published a handful of essays in the still-somewhat-alternative *Village Voice,* where I could assume a readership with political sensibilities at least broadly similar to mine. Then in the last months of that year I had essays accepted by more "mainstream" publications like *Harper's* and the *New Yorker*—magazines that, unlike the *Voice,* are printed on glossy paper and are indexed in the *Reader's Guide to Periodical Literature.* In each case, I wondered whether, in order to appear in such venues, I would have to do some measure of selling out—as an academic, as a fire-breathing progressive, whatever. And this was a weird feeling for someone who's argued, like Gerald Graff, in *favor* of our co-optation by the "mainstream."

So I think it would be fatuous of me to stress the necessity of doing "public intellectual" writing, as if this were a matter of simple volition. More important, I want to say that my experience of writing for those magazines, limited as it is, did indeed involve some kind of selling out—a kind of selling out that has impelled me to think anew about the potential and actual relations between cultural studies intellectuals in the academy and the discourses of cultural politics and social policy currently popular in the United States. What is of greatest moment to me in this regard, what seems to me the signal crisis in these actual and potential relations, concerns the idea of the national "public" itself, and the status of this idea in an era when public housing, public education, public health, public ownership, public welfare, and public funding for television, the arts, and the humanities have all come under savage attack from our so-called national leaders.

In this climate it is deeply disturbing to me how successfully both wings of the New Right, the economic-libertarian and the cultural-fascist wings, have been able to attack "the public" in the name of the people; this seems to me a new and virulent strain of authoritarian populism the antidote to which we have not yet been able to imagine. But it is also deeply disturbing to me that there is so much skepticism and outright disdain, among the academic Left, directed at the proposal that intellectual work in cultural studies should seek to have an impact on the mundane and quotidian world of public policy. When American cultural studies theorizes "the everyday," it appears, "the everyday" does not always involve school breakfast programs, disability law, or the minimum wage. I'll get back to this line of argument in a moment; first, I want to explain how I have had to sell out, and then I want to say a few words about the avant-garde tradition that still influences many American left intellectuals, a tradition in which negotiating with the state, with Ideological State Apparatuses (ISAs), or with mainstream print media in civil society is considered not one of the obligations of citizenship but a form of capitulation to capitalism.

My encounter with *Harper's* is not especially germane to this chapter, except that insofar as it broached the subject of human disabilities and the discourse of genetic foundationalism (it's about my son James, and eventually formed the basis of my first "crossover" book, *Life As We Know It*), it did attempt to intervene in the neo-eugenicist debates surrounding *The Bell Curve;* I felt as if I were aiming a peashooter at the Charles Murray think tank. But it's worth noting that when I sent the essay to *Harper's* at the suggestion of a friend, the first thing I was told was that it would have to be cut in half, and most of the citations would have to go. My editor, a smart and judicious young woman, informed me that she had an extremely stringent criterion for magazine essays: ideally, she said, you should be done reading them before you realize you've begun. She is not averse, she explained, to sinuous narratives, patient excursuses, or carefully modulated philosophical deliberations— except in magazine essays, which should seem so effortlessly written as to betoken an equally effortless task of reading. I had some reason to worry whether my essay would pass that kind of muster, and whether I should ask it to; the article was originally written for an academic collection of essays, and was full of excursuses and deliberations on evolution, meliorism, and the survival value of intelligence. My opening sentence was "Down syndrome does not exist; what exists are the practices by which we know and produce Down syndrome"—a sentence I thought anyone familiar with American cultural studies would recognize as an allusion to Douglas Crimp's citation of François Delaporte at the opening of *AIDS: Cultural Analysis/Cultural Activism.* Needless to say, that sentence had no allusive force whatsoever for the hypothetical or actual readers of *Harper's,* and I was advised to cut the opening few pages on social constructionism and replace them with the story of my son's birth.

I agreed to almost all of the most severe editorial suggestions, and the result was actually a much tighter, much less self-indulgent essay, one whose most important theoretical points about genetics and language I was forced to embed *in* my narrative rather than appending *to* the narrative. But oddly enough, I found myself accused of selling out nonetheless, by a colleague who told me that he liked the piece except for what he called the sellout to liberal humanism in the final two paragraphs. For him, it was as if I had had to attach those paragraphs before *Harper's* would accept the essay. As it happens, those two paragraphs had been in the essay right from the start; I had imagined them as attempts to negotiate the work of feminist Habermasians with the work of psycholinguist Steven Pinker (and I had had a long conversation with my editor about what this negotiation entailed).[1] Accordingly, I told my colleague that matters were even worse than he thought: I actually *believe* my conclusion that we should sign unto others as we would have them sign unto us. That's no sellout, I said, that's just me. The only "sellout" was leading the essay with the story of Janet's labor on the day James was born.

The *New Yorker* piece presented me with an altogether different, and more dangerous, set of challenges. My assignment was to write a review essay on the emergence of black intellectuals as public intellectuals, focusing specifically on Cornel West, bell hooks, Michael Eric Dyson, Thomas Sowell, and Derrick Bell. (That lineup was hashed out over time; the first draft of my essay also included figures like Henry Louis

CULTURAL CRITICISM AND THE POLITICS OF SELLING OUT

Gates, Patricia Williams, and Jerry Watts.) Taking the New York intellectuals of the 1940s and 1950s as my point of contrast (Daniel Bell, Irving Howe, Philip Rahv, Lionel Trilling), I decided that my job was basically to file a partisan review. And since West and hooks were already academic "celebrities," I didn't have to worry about making their work "accessible" to the readers of the *New Yorker;* I could assume some general knowledge of their work, and proceed to criticize it on its merits, devoting very little time to paraphrase.[2] My editor at the *New Yorker,* Henry Finder (also the managing editor of *Transition*), was sympathetic enough to this approach, but turned back my first draft by asking (a) for more detailed argument on the work of Cornel West, and (b) whether I couldn't be more skeptical of the academic Left in general than I had been.

At this point I got nervous, and not merely because of the possible racial politics involved. For one thing, I was familiar with Eric Lott's stinging review of West's *Race Matters* in *Social Text,* and although I had wide areas of agreement with Eric's essay, particularly with the charge that West comes close to pathologizing the black underclass, I most certainly did not want to adopt Eric's strategy of attempting to represent the forces of vitality and resistance in the black underclass more fairly than West. It was a risky enough strategy when Eric used it, and I did not like contemplating the prospect of insisting, from the study of my central Illinois home, that the black urban subaltern could speak in ways Cornel West wasn't hearing—as if this should become the method of choice for white critics looking to trump their black colleagues. But more than this, I wondered how I could convey some of my own skepticism about the political effectiveness of the academic Left without seeming to dump it all on the heads of people like West, Dyson, and hooks. And, of course, I asked myself whether, in going along with the request to devise a more skeptical conclusion to the review, I was simply selling out. My ambivalence became all the more violent when I opened the *Chronicle of Higher Education* of December 14, 1994, right in the middle of rewrite number twenty, to find that the "Hot Type" column included a précis of Eric Lott's *Social Text* essay, which began, "Cornel West is in danger of selling out, says Eric Lott" ("Hot Type" 10). I had read Eric's essay in manuscript some months before, but I had not anticipated so tangled a course as this: what are the cultural politics involved, I wondered, in the possibility of selling out by joining the growing chorus of voices accusing Cornel West of selling out?

This was not merely a matter of academic politics, of who gets to say what about whom and who will think what of whom as a result. Rather, it was a question involving what Edward Said has called "representations of the intellectual." Because I had been asked to discuss these figures as public intellectuals, I had decided to base my claims for their work on the fact that they possess a constituency, largely but not wholly African American in composition, for and to whom they regularly speak; and I meant to juxtapose that sense of constituency to the self-conscious cosmopolitanism of the New York intellectuals of the 1930s and 1940s—a cosmopolitanism that helped establish them *as* intellectuals, but which, regrettably, sometimes left them unable or unwilling to speak as *Jewish* intellectuals.[3] Therefore, it seemed to me, the way to discuss the functions of new black intelligentsia, for better *and* for worse, was to discuss the question of intellectuals' constituencies. For theorists like Julien Benda, as for a long tradition of avant-garde intellectuals from Henri de Saint-Simon right through to Mas'ud Zavarzadeh, an intellectual who has a constituency is by definition an intellectual who has sold out: the intellectual is he (almost always he) who transcends all affiliations. As Said writes, such intellectuals "have to be thoroughgoing individuals with powerful personalities and, above all, they have to be in a state of almost permanent opposition to the status quo: for all these reasons Benda's intellectuals are inevitably a small, highly visible group of men—he never includes women—whose stentorian voices and indelicate imprecations are hurled at humankind from on high" (7).

I fancied that my solution to the problem of how to be suitably skeptical of public intellectuals considerably more "public" than myself had a certain attractive recursivity to it: I would sell out by broaching the subject of whether intellectuals with constituencies have sold out to the extent that they "represent" a public at all. And since the intellectuals in question are black, I could broach this question simply by attending to the ways they themselves broach the question, for it is a crucial issue for them: "because non-black audiences are still the ones that have the power to put black artists at the top of the charts," I wrote,

> African-American intellectuals' uneasiness about black commercial and professional success stems in part from the long-standing fear that "crossing over" must entail selling out. It's what leads to [bell] hooks' attack on [Spike Lee's *Malcolm X*] [which, she'd written, "cannot be revolutionary and generate wealth at the same time"]—the unstated suspicion that any critical or commercial success with white audiences is, de facto, political failure. (77)

And I concluded that "if black intellectuals are legitimated by their sense of a constituency, they're hamstrung by it, too: they can be charged with betraying that constituency as easily as they can be credited with representing it" (77–78).

All well and good. But as fortune would have it, I had been given a deadline of November 11 for the *New Yorker* essay, which meant that I was writing it just as the election returns of November 8 began to be narrated as an epochal moment in postwar American politics, the "romp to the right" (as *USA Today* called it) that finally freed the American people from the tyranny of liberal Washington. I therefore decided that it was time for me to unburden myself of some of my latest obsessions about cultural studies, regardless of whether they were apposite to the task at hand: to wit, I would take Thomas Sowell's *Race and Culture: A World View* as the occasion to note that "where the left tends to address itself to culture, the right—even when it takes up the topic of 'culture'—tends to address itself to policy" (79). This phenomenon is particularly remarkable in Sowell's book precisely because it claims, at the outset, not to have any designs on practical policy making; the rest of the book, however, mounts a case (and stop me if you've heard this one before) against the evils of the custodial welfare state and the doctrines of cultural relativism in the social sciences, which apparently sustain the welfare state. My concern was, and is, that right-wing intellectuals from Charles Murray to William Kristol to James Q. Wilson seem never to open their mouths but to articulate their versions of "cultural studies" to the exigencies of policy making, whereas cultural studies theorists on the Left often express outright disdain for the policy implications of their work, as did Andrew Ross in an interview in *Lingua Franca* when he dismissed the task of addressing policy as "a little too easy" (60).

This then has been my fixation since the elections of 1994: configuring the relations among American cultural studies, the latest policy initiatives of the New Right, and the discourse of the public intellectual. I want to argue that cultural studies, if it is going to be anything more than just one more intellectual paradigm for the reading of literary and cultural texts, must direct its attention to the local and national machinery of public policy. But I also want to argue that doing academic cultural studies, as currently constituted, is not enough — just as writing a few essays here and there for mass-market magazines is not enough. Teaching and writing are two important ways of being public, and we need to say so whenever we are publicly accused of being insufficiently public; but what I want to call for is a practice of cultural studies that articulates the theoretical and critical work of the so-called public intellectual to the movements of public policy.

One advantage to voicing these concerns in the *New Yorker,* I thought, was that it would allow me to chip away at the legend of the New York intellectuals, by pointing out how dismal was their own record in this regard. I beg your indulgence for one more quote from that essay, one that summarizes my concerns about cultural politics while refusing the nostalgic fantasy that Howe and Trilling were giants who walked the land:

> watching the American left redefine the terrain of cultural politics while practical political positions to the left of Bill Clinton disappear from the map, one begins to wonder if there isn't a sense, even in the work of the most prolific and capable black public intellectuals, that cultural politics is a kind of compensation for practical politics — more satisfying, more supple, more susceptible to sheer intellectual virtuosity, because it involves neither revenues nor statutes. Not that the celebrated New York intellectuals of yesteryear held any clear advantage in the realm of practical politics: Lionel Trilling had no hand in Truman's

Far East policies, nor did Philip Rahv help enforce Brown v. Board of Education. (79–80)

The word "yesteryear" in this last sentence is crucial to the point—since, after all, we now live with the legacy of the New York intellectuals in the form of William Kristol, and it is Norman Podhoretz's magazine, *Commentary,* that has recently defended the brutal beating and murder of gay men on the grounds that it discourages "waverers" from indulging in the gay lifestyle.[4] But I don't want merely to remark on the sorry transformation of the New York intellectuals into the neoconservatives; I want also to introduce into our discussions of the "public intellectual" the overdue recognition that the New York intellectuals were often the worst kind of armchair quarterbacks and fence-sitters, "activists" whose only activism consisted of essays in *Dissent* or *Partisan Review.* Time and again, when crucial social issues were on the table, the New Yorkers elected to pass, and when it came to taking stands on the Vietnam War, on school desegregation and decentralization, on the women's movement, many of the so-called public intellectuals of the 1950s and 1960s compiled a deplorable record.

My point in rattling these old bones is not simply to challenge the lionization of the New York intellectuals, but to raise a question I think is crucial to our own moment, in which the challenge of the would-be public intellectual is precisely to revivify our conception of "the public." It is remarkable, to say the least, that even moderate Republicans like New Jersey governor Christine Whitman have so successfully elided the rhetoric of public ownership with the rhetoric of government control, as when she suggested, apropos of funding for the Corporation for Public Broadcasting, that the notion of government ownership of media went out with the passing of *Pravda.* This bizarre logic, I note, has given us not merely a new form of discourse, in which eliminating the National Endowment for the Arts is construed as a means of liberating artists from the shackles of government funding; it has also given us a whole raft of toxic new policy initiatives, from the repeal of laws ensuring environmental protection and workplace safety to the insurance industry boondoggle known as "tort reform." And one of the reasons for that, of course, is that conservative public intellectuals see it as integral to their enterprise to undermine the idea of the public in the realm of public policy—which is why the aging Hitler Youth duo of Peter Collier and David Horowitz make sure to send their interventions in cultural politics (namely, the tabloids *Comment* and *Heterodoxy*) to the offices of elected officials who will aid them in their noble quest to gut the public.

Nonetheless, I knew—and know—full well the trouble I invite by suggesting a division between "cultural politics" and "public policy." With the American Left in such woeful disarray, it is all too often that we hear about how the "academic Left" or the "cultural Left" is obsessed with its readings of Madonna, its barbaric jargon, and its byzantine rituals of theoretical purification. And sure enough, among the first "progressive" responses to the Republican sweep was Michael Tomasky's cover article for the *Village Voice,* blaming the academic Left for the rise of Newt:

> so we sit around debating the canon at a handful of elite universities and arguing over Fish's and Jameson's influence on the academy, while the vast majority of working-class young people in America (a) will never read the canon, however you choose to define it and whatever you wish to in- or exclude, (b) will think Fish and Jameson stand for a dinner of carp and Irish whiskey (and be little the worse off for thinking it, incidentally), (c) will take very few literature courses, and (d) will be working like hell to save the money to pay their tuitions at a two-year college or perhaps a land-grant university. (19)

He then goes on to cite who else but Russell Jacoby. Tomasky was promptly rebuked in the pages of his own paper, chiefly by Ellen Willis, but the charge stung precisely because we have heard it so often before, particularly from self-described liberals such as David Bromwich, Daniel Harris, Russell Jacoby, and Richard Rorty. For that matter, the charge that the academic Left is too self-absorbed is not wholly without merit; it is simply a different kind of claim from the charge that the academic Left is marginal to national politics, or that the academic Left is responsible for the 104th or 105th Congress. You can blame people for their self-absorption, but you cannot necessarily blame them for being marginal, and analyses like Tomasky's (which he has since expanded into a book, *Left for Dead*) tend to confuse this issue, provoking nothing more productive than defensiveness and academic turf-guarding in response. Likewise, even if I were to enumerate and denounce every kind of academic self-absorption and ineffectuality I know of (and this would take some time), I would still not have shown that any of it was to blame for the latest conservative turn in national politics—a phenomenon we might more plausibly attribute to the malfeasances of elected Democrats and the terrifying mass mobilization of the Christian Coalition.

Whatever my own qualms about the academic Left, then, I do not want to be saddled with positions I have not taken: I do not claim that cultural politics isn't a "real" politics, nor would I claim that struggles over popular culture are unrelated to struggles over public policy. To quote an influential theorist I've been reading lately, "it is simply not possible to refocus this nation's public policy debate through electoral politics alone" (87); moreover, "the left has been very successful because it understands the importance of the culture—of framing the debate and influencing the way people think about problems" (88). For those of you who don't recognize the prose style, that was Rush Limbaugh. I think he's got a point, and I think there are many reasons the academic Left devotes so much of its attention to culture. Not least important of these, I submit, is the fact that most citizens of the United States devote more of their attention to culture than to politics; it is no exaggeration to say that most Americans live their relation to the political by way of the cultural, as was amply demonstrated by 1994's public debates over films like *Forrest Gump* and *The Lion King.* And, as I've argued elsewhere, cultural criticism is ubiquitous on the American political landscape, particularly the kind that proceeds from figures like Michael Medved, Cal Thomas, Rush Limbaugh, William Bennett, and Christina Hoff Sommers, not to mention Peter Collier and David Horowitz. Finally, there's the eerie fact that the realm of popular culture often seems to offer wiser and more bracing analyses of post-Fordism and the crisis of the American worker than anything ordinary people can find in the political realm. My favorite example here comes from *The Simpsons,* in an

227

episode in which Homer and a coworker are representing the company in the state capital, and decide to order room service at their hotel and put it on the company tab. No sooner do they do so than a red buzzer goes off hundreds of miles away in the office of Mr. Burns, the CEO, whereupon Smithers, Burns's assistant and sycophant, remarks, "Someone's ordering room service, sir." Burns then orders Smithers to release the winged monkeys, and in a brilliant citation of *The Wizard of Oz,* Burns cackles, "Fly! Fly, my pretties!" But the monkeys crash to the ground almost immediately, provoking Burns to mutter that the program still needs more research. I want to point out that my then seven-year-old son enjoyed this surreal scene every bit as much as I did, which leads me to believe that when it comes to depictions of the post-Fordist economy in which obscenely wealthy CEOs cook up draconian schemes for employee policing, we can say of *The Simpsons* what Augustine said of the Bible: that its surface attracts us like children and yet its depths are stupendous, rendering its meaning copiously in so few words.

So if it is true that cultural politics sometimes seems like a compensation for practical politics (and it is certainly true in my own life), perhaps this is so because Americans often appear smarter as consumers than as political agents; at the very least, we certainly tend to understand ourselves more readily as consumers than as political agents, which is one reason, as James Carville has often remarked, so little of the American electorate has any substantial understanding of political issues that do not directly affect their disposable income.

Even in the political arena bounded by the state, however, there is no clear distinction between cultural politics and public policy, and therefore no clear way of determining whether or not "politicizing" an issue is a responsible or irresponsible thing to do. When, in the spring and summer of 1994, the Congressional Black Caucus sought to "politicize" the president's crime bill by demanding that it retain a provision on racial justice and the death penalty, they were undoubtedly speaking cultural politics to policy in a manner that should remind us how tenuous is the distinction between the two. But even here we can find two senses of the word "political": the narrow sense in which the caucus informed Clinton that he might not have the votes to pass the bill unless he addressed their concerns, and the larger sense in which the caucus was trying to redirect the subject of race in our national discussions of crime and punishment, such that African Americans would be seen not as potential felons but as citizens targeted by a police apparatus in which they are many times more likely than their fellow white Americans to be stopped by the highway patrol, to be scrutinized in retail stores, and to be given the death penalty for violent crimes. Racial justice, in other words, is not solely a matter of public policy. And yet the cultural politics of racial justice have had to address the vicissitudes of public policy, ever more urgently since the insane Supreme Court decision of *McClesky v. Kemp* (1987), which held that statistics concerning race and executions, regardless of the weight or clarity of the statistics, were immaterial to challenges under the Eighth Amendment.

Now, I know that in making this argument I'm running the risk of preaching to the converted, bringing coals to Newcastle, and doing any number of similarly pointless things. I don't imagine that very many of my actual or hypothetical readers opposed that provision of the president's original crime bill on the grounds that it would create an unacceptable "quota system" for lethal injections.[5] I adduce this example only because I want to walk this line carefully, and to insist not only on the difference between cultural politics and policy but also on their inevitable entanglement. What I'm saying in this regard has in fact been said before, most notably by Tony Bennett in an essay entitled "Putting Policy into Cultural Studies." But it is symptomatic of our uncertainty about the politics of intellectual work, I think, that Bennett's essay would have gotten so hostile a reception from so political a theorist as Fredric Jameson, who wrote in his *Social Text* review of the Routledge *Cultural Studies* collection that Bennett does not "seem to realize how obscene American left readers are likely to find his proposals on 'talking to and working with what used to be called the ISAs rather than writing them off from the outset and then, in a self-fulfilling prophecy, criticizing them again when they seem to affirm one's direst functionalist predictions' " (29). It is when I read passages such as this that I begin to fear the creation of an academic Left whose only function is to analyze and interpret the formation of the hegemonies that are actually *being* formed by our counterparts on the Right; I fear an intellectual regime in which cultural studies is nothing more than a parasitic kind of color commentator on the new authoritarian populism of the Age of Gingrich, too busy explaining the rise of the postmodern eugenicist-libertarian-cybernetic-fundamentalist Right to be of any use in actually opposing it. If you look at Limbaugh's *See, I Told You So* you'll see something of a looking-glass version of what I'm talking about; the passage I quoted earlier, in fact, appears in a chapter in which Limbaugh laments that all the liberals are reading Gramsci (would that this were so) and are engaging in a "war of position" against traditional American values. What Limbaugh cautions his readers against, needlessly but always strategically, is a world in which the Left has all the tools to wage wars of position and the Right is only belatedly trying to understand how the Left so thoroughly dominates the nation; and I don't think it's paranoid or defeatist to suppose that at the moment, the reverse is much more nearly the case in the United States.

I want to remark briefly on the irony of broaching this issue by way of discussing black public intellectuals. There seemed to me—and there still seems to me—something grossly unfair about discussing the disjunction between left intellectuals and public policy by focusing on people like bell hooks or Cornel West or Derrick Bell, who do cultural politics that often have everything to do with public policy. In the *New Yorker,* for example, I called attention to a passage in Michael Eric Dyson's *Making Malcolm: The Myth and Meaning of Malcolm X* in which Dyson relates a compliment he received from a young black admirer. He had just testified at a Senate judiciary subcommittee hearing on gangsta rap, and he had quoted Snoop Doggy Dog verbatim from memory, prompting an observer to tell him, "for a guy your age, you really can flow" (xxi). In the context of Dyson's book, the story concerns Dyson's own relation to youth culture and to the discourse of black urban masculinity in crisis; but for want of a better alternative, I took Dyson's story as the occasion to remark that "for cultural critics, the danger of popular acclaim is that it can tempt them to pay more attention to the responses of young admirers than to the deliberations of Senate subcommittees" (79). Dyson, I think, would have every reason to ask who's zooming who here; and should he ask, I would have to admit that I myself have never been asked to testify before the Senate on any subject whatsoever. But Dyson's presence in Washington only underscores my point: when the Senate judiciary subcommittee opened its hearings on violence and misogyny in gangsta rap, Carol Moseley-Braun called on academic critics such as Dyson and Tricia Rose, who showed up to tell the Senate they were asking all the wrong questions. What were Dyson and Rose doing, I ask you, but talking—and talking back—to the ISAs? I stand by my point regardless of how ill-gotten it may have been in regard to Dyson—and what's more, I wish his book *Making Malcolm had* said more about Dyson's role in those hearings. Dyson's work on Malcolm is scrupulous and, in its way, as relentless as is bell hooks in showing that the rhetoric of black nationalism renders a black subjectivity that is always already a black *male* subjectivity. This too has something to do with the politics of lethal injections, as Ice Cube's album of the same name suggests, and it cues us once again to the myriad connections between cultural and practical politics. But to repeat the point one last time, it's one thing to make the entirely necessary argument that black nationalism foregrounds a masculinist politics and a male subjectivity; it's another thing to translate that point into a political initiative that will realize some of its many implications for the disposition of revenues and statutes.

I can imagine three salient objections to what I've argued thus far, and I want to mention them partly because I want to agree with them ahead of time. The first is that the connections between intellectual work and the policy implications thereof are too often drawn sloppily or arbitrarily; accustomed as we are to unfolding arguments with care and rigor, it seems superfluous or even humiliating to have to tack on some kind of policy prescription once we've finished historicizing the nationalist subject. The second is that the connections between cultural politics and public policy are too often drawn at the expense of the former. This objection takes many forms, from Andrew Ross's complaint that the aim of intellectual work should be to change consciousness rather than to enact statutes, to Edward Said's warning that the intellectual who speaks truth to policy may wind up as the tool of the political apparatus he or she seeks to operate, to my own sense that cultural studies can be more easily compromised to fit the demands of public policy than can public policy be reimagined so as to accommodate the utopian yearnings of cultural studies. And the third objection, which no doubt occurred to some readers once they'd read my first few paragraphs, is that right-wing intellectuals are *by definition* closer to the sources of national political and economic power than any of the liberals and leftists I've mentioned; and it is only in this sense, I think, that we can make sense of Jameson's repudiation of Tony Bennett's proposals as "obscene."

Because there's nothing we can do about this third objection, other than hoping that more progressive organizations will see the need and have the capacity for massive fundraising at the national level, I'd like to address the first two for a moment. The link between theory and policy often is capricious and ill-conceived: as Jacob Weisberg pointed out in his review of *The Bell Curve,* you could accept every piece of Herrnstein and Murray's work, the shoddy statistics, the impoverished measurement of "intelligence" by the g factor, the unwarranted confidence in our ability to define race, even the questionable use of "research" sponsored by neo-Nazi groups, and you could still come to the opposite conclusions from those drawn by Herrnstein and Murray. That is, you could posit a permanent and ineradicable intelligence differential based on "race," whatever that is, and argue for the maintenance and expansion of the welfare state on precisely those grounds. Conversely, it would not be hard to imagine an argument for dismantling what remains of the welfare state on the reasoning that there is no correlation whatsoever between race and intelligence, and that affirmative action programs are therefore interfering with the "natural" distribution of intellectual talents.

I was rather forcibly brought up against this conundrum in my reading of Thomas Sowell's *Race and Culture,* a book I'd like to offer as something of a model of right-wing global cultural studies. Knowing that in the *New Yorker* I would have no more than two paragraphs in which to dispose of Sowell, I decided to concentrate on his "Race and Intelligence" chapter—again, partly because this is one of my signal obsessions at the moment, and partly because I thought it would be rhetorically effective. Sowell's data happened to overlap with the data I had learned about from another source, namely, James Trent's social-constructionist history of mental retardation in the United States, *In-*

venting the Feeble Mind (which documents the symbiotic rise of eugenics and intelligence testing in the first decades of this century, and documents the policies of sterilization and institutionalization that began in the United States in the 1920s and were then expanded to great effect by Nazi Germany in the 1930s and 1940s). So, drawing on Trent together with Anthony Appiah, I thought I could make the point clearly enough: once you start historicizing and globalizing comparative studies of race and intelligence, you give the game away, by showing that intelligence differentials are in fact social and cultural rather than biological.

But since Sowell explicitly rules out the biological explanation of intelligence elsewhere in the book, I would have to make the point much more carefully than I thought—since Sowell's argument itself is more careful than I had at first imagined. Here then is what I finally devised, with the help of a remarkable *New Yorker* fact-checker by the name of Blake Eskin:

> "For many practical purposes," he concludes, "it makes no differ-
> ence whether poor performances in abstract thinking are due to
> neglect or to lack of capacity." But this is no conclusion at all,
> since if groups' differences *are* attributable to "neglect"—or
> worse, to active discrimination—then Sowell has inadvertently
> demonstrated the necessity for precisely the kind of ambitious
> social programs his career has been dedicated to attacking. (78)

I had fondly hoped that if I exposed this contradiction in a major magazine, the exposure would have some effect on something. But I have to admit what I've only alluded to above: despite my best efforts to analyze, historicize, deconstruct, and redefine the arguments of my opponents, I actually have not had all that much impact on national policy myself. The 104th and 105th Congresses have gone about the business of dismantling social programs despite the fact that one radical conservative claims natural, biological sanction for doing so, and another radical conservative explicitly rejects that reasoning yet comes to the same conclusion. It would seem, then, that right-wing intellectuals will somehow come to the conclusion that the unregulated "free market" provides the best of all possible worlds regardless of whether they find justification for

233

this belief in nature or merely in culture. Even when Sowell explicitly disagrees with Murray's premises, then, he nevertheless manages to agree with him on the policy tip. Borrowing a page from Andrew Ross, I like to think about this phenomenon in terms of the weather: Charles Murray wakes up, predicts sunny skies and a high in the sixties, and concludes that today would be a good day to eradicate the notion of racial justice; Thomas Sowell wakes up, predicts rain turning into sleet by nightfall, and concludes that today would be a good day to eradicate the notion of racial justice. It is no mystery why we should be skeptical about the prospect—or the potential value—of putting policy into cultural studies.

Nonetheless, my worry is not that academic progressives will construe cultural politics as something wholly other than practical politics; my worry is that we will tend to conflate them. In other words, I really don't see any danger, at the moment, that the cultural Left will decide that popular culture is not a proper location for political struggle, or that debates over race, ethnicity, clothing, cuisine, music, science, and technology have no manifest "political" content. On the contrary, I see plenty of danger that we will underestimate—or, worse, ignore—the difference between theoretical work on such subjects and the practical political effects such work can have for the people we're talking *about* if not necessarily talking *to*.

One recent instance of the disjunction between cultural studies and public policy stands out especially in my mind, partly because it's drawn from a book whose theoretical sweep and ambition I admire and cannot hope to emulate, and partly because the book in question explicitly sets out to understand and contest the new conservatism in American life. At the end of *We Gotta Get out of This Place,* Larry Grossberg alludes to the battles over "political correctness" that were erupting just as he was finishing his manuscript, and he castigates the Left's responses to PC, including my own, for how they "manage to avoid most of the issues and absolve themselves of all blame" (430). Wondering what issues we had avoided and what blame we should have accepted, I read further that "some people have taken extreme, absurd positions which should be criticized" (an interesting injunction, since Grossberg does not cite specific persons or positions) and that "the Left has to recognize the truth in the Right's accusations: e.g., affirmative action in universities has not solved the problem and has created new ones" (383).

I want to remark on this notion that in the early months of 1991, as the PC attacks swept across the American media, we on the Left should have taken some of the blame for the failures of affirmative action. I don't think Grossberg is wrong to point to these problems; I just think that one could not plausibly single out affirmative action as terrain for concession to the New Right if one were paying even the faintest attention to the policy disputes of 1991. For it was then that the Civil Rights Act of 1991 was being bandied about in Congress, an act that specifically sought to hold employers to the standard that any job qualifications that screen out minorities and women must be related to a person's ability to perform the job. In 1989, the Supreme Court had held, in one of its most perverse opinions, *Wards Cove v. Atonio,* that American courts had never established such a "job performance standard," and the Bush administration, led by C. Boyden Gray and Dick Thornburgh, was pushing for a weaker standard, requiring companies to show only that their hiring practices serve "legitimate employment goals" and requiring employees to identify the specific employment practice that brought about the alleged discrimination. (As policy analysts pointed out, this would allow employers to defend discriminatory practices on such vague grounds as corporate image, customer preference, or employee morale.) The bankruptcy of the Bush/Gray position, I should note, was quickly exposed by the NAACP Legal Defense and Education Fund, which showed that in 217 of 225 cases since 1971, the courts had indeed used the "job performance standard" the Bush administration and the Rehnquist Court claimed not to exist.[6]

Now, let me remark on one important consequence of this somewhat arcane policy dispute. The most prominent and vocal Republican supporter of the Civil Rights Act of 1991 was Senator John Danforth of Missouri. Bush eventually signed the bill, but only after Danforth had served as Clarence Thomas's point man and chief defender in the Senate in an unofficial quid pro quo for his (Bush's) capitulation to what right-wing intellectuals to this day fraudulently call a "quota bill." In the larger agenda of the Bush Administration, in other words, passage of the bill would be paid for by the appointment of Clarence Thomas to the Supreme Court. No better example could be devised, I submit, of the profound interdependence of cultural politics and practical policy making; and in this case, I can think of no better example of the cultural Left's often lamentable inattention to policy matters than Grossberg's suggestion that in the PC debates of 1991, we should have acknowledged the "truth" in the Right's attacks on affirmative action.

I am not suggesting that we all be called to account for the policy implications of our work, nor am I demanding that we each append to our next book or article a little "policy epilogue" that spells out the practical steps that follow from our analysis of Madonna, Malcolm, Macherey, or nationalism, imperialism, or gender and the public sphere. I am simply asking that we be attentive to the ways cultural studies might conceivably be of interest to, or intersect with the work of, theorists and political agents more directly involved with the policy machinery of the state. And this is why, as I look over the textual record, I wound up unable to write the essay I'd originally planned: because to make the kind of case I'm making in *this* essay is to call out the limitations in my own work. To date, my conception of the so-called public intellectual has relied almost exclusively on the very model I criticize in the New York intellectuals, the same model that still dominates most of our discussion of the subject: the intellectual who engages with what Peter Uwe Hohendahl, following Habermas, calls the "literary public sphere." My conception of selling out, therefore, was confined solely to the question of how American cultural studies could best represent itself in the mainstream press. I now want to suggest that there is at least as important a difference between the literary public sphere and the public policy sphere as the difference between cultural politics and public policy, and that most cultural studies intellectuals, myself most assuredly included, have not yet begun to think seriously about how best to negotiate that difference.

To admit that difference and to seek to negotiate it would be to sell out, and no one has said so more emphatically, of late, than Mas'ud Zavarzadeh. In one of his most recent attacks on the ludic American

academy, Zavarzadeh writes that "the various tendencies of ludic populism can perhaps best be outlined by examining the emerging figure of
the post-al 'public intellectual' " (106). Zavarzadeh goes on to insist
that "the bourgeois 'public intellectual'—in the name of democratizing
knowledge—perpetuates the ignorance of the people and deepens their
dependence on the knowledge industry" (107). The heart of his critique
goes to the heart of my argument:

> the credibility of the bourgeois "public intellectual" is established
> by his/her "activism," which is, itself, an "affirmation" of the
> system by accepting (affirming) its rules and playing inside the
> system according to the rules of reform. . . . The "public intellec-
> tual" is a figure invented to combine this deep anti-intellectu-
> alism and counter-revolutionary affirmation of the commonsense
> with reformist localism. (107)

To most of Zavarzadeh's indictment I have to plead guilty as charged:
I am committed to playing inside the system according to the rules of
reform—though I've argued as often as I can that there's all the difference in the world between weak and strong reformism (and, it should go
without saying, I am anti-intellectual and counterrevolutionary to boot).
But I will differ with Zavarzadeh in one respect: I do not want to
fetishize the local. On the contrary, it seems to me that what we need
most desperately in the wake of the elections of 1994 (from which we
have still not recovered) are new discourses of national identity, new
discourses of national unity. The academic Left normally does not even
contemplate the possibility of such discourses except as expressions of
conservative, nostalgic, cryptofascist fantasy. But while we've been making the case against imposing or presuming a common American culture,
the New Right has worked assiduously to destroy the material foundations of what can at least potentially sustain us as a common *society.*
That's why their attacks on the realm of the public are so important, and
why it is so important that we reclaim and rejuvenate "the public" in the
name of the people. It may seem comic or tragic that the New Right so
painstakingly undermines our common social foundations, like public
schools and social services, at the same time that it makes a cottage

industry of screeds lamenting our loss of a common culture or a common morality. But I think we should understand these phenomena as complementary symptoms of the same general development: as Congress undoes the social-welfare powers of the federal government by parceling out social services in "block grants" to the states, thereby reminding us that the purpose of the so-called New Federalism of the 1980s was precisely to Balkanize our national laws pertaining to taxation, abortion, welfare, schooling, employment, and housing, we will undoubtedly hear all the more empty, idealist nationalist rhetoric about the need to affirm traditional values and bring back a common sense of shame—and we will hear this, needless to say, from people whose political behavior is itself nothing other than shameful. As Samuel Luttwak has written in a brief essay cheerily entitled "Why Fascism Is the Wave of the Future,"

> It is only mildly amusing that nowadays the standard Republican/Tory dinner speech is a two-part affair, in which part one celebrates the virtues of unimpeded competition and dynamic structural change, while part two mourns the decline of the family and community "values" that were eroded precisely by the forces commended in part one. (6)

It is in this moment of crisis, then, that cultural studies intellectuals need to imagine a language wherein national identity and American cultural politics are mobilized not as they were in California, when a rainbow coalition of hyphenated Americans got together to pass Proposition 187, but wherein "patriotism" is redefined as that sentiment that prevents us from letting our fellow citizens starve, beg, or go homeless. The Right knows that its rewards for detaching a sense of national identity from a sense of public citizenship can be great indeed: as long as the ideological construction of "American identity" has nothing to do with access to housing, health care, employment, or basic nutrition, our leading plutocrats can continue to reap the benefits of the industrialized world's most inequitable tax system while our fellow citizens live with the industrialized world's highest rates of child mortality and childhood poverty. And if we do not imagine an alternative conception of national identity, I fear that our federal government's fiscal powers will be further weakened as its police powers are steadily expanded—such that fascism, to borrow Luttwak's phrase, will indeed be the wave of the future.

I realize that I run the risk of reaffirming the very discursive formation I seek to undermine; in focusing on national identity I may be simply proposing nationalism as a refuge from capitalism, and thereby hastening the rise of the repressive, punitive discourses of national identity we see at work in California. I must admit I am not sure what I think about this: for Luttwak, it is the globalization of capital rather than the resurgence of nationalism that threatens to precipitate the return of fascism, and in an era when the most vocal opponents of GATT and NAFTA have been Patrick Buchanan, Ross Perot, and Ralph Nader, I have to say I do not know what the relations between nationalism and capitalism may be in the future, or indeed what they may be in the present. I know the danger here is that of proposing reactional countermoves, concocting counterdiscourses, and ultimately playing into the hands of the ISAs. But that is a danger I believe we must court; and I believe we have nothing to lose in doing so—except our theoretical purity.

I want to close with one parable about that danger. I was perusing the Down syndrome discussion group on the Internet in the winter of 1994–95 when I saw a posting from a self-described conservative Republican that advised parents of disabled children how to deal with the newly Republican Congress, particularly if they'd just lost a liberal Democrat to a conservative Republican. Since the abolition of unfunded federal mandates has the potential to eviscerate the Americans with Disabilities Act, and since on the state level we had indeed lost a progressive activist to a local right-wing machine politician, my spouse and I read this posting with great interest. It advised us not to talk about "rights and entitlements," but to emphasize instead our child's potential for "self-sufficiency"; and not to talk about "justice" or "fairness," since life is inherently unfair, but to insist on the value-neutral principle of equality before the law. Much as we hated to admit it, we thought this was good, well-intentioned, pragmatic advice, and we were briefly grateful for the lesson in how to tailor our political convictions to the purpose of maintaining our son's social services. *Briefly* grateful, that is, until we read a brilliant posting from a woman named Janet Curtis, who, identifying herself as "nothing more than a poor housewife," fired back a reply that said, in effect, how *dare* you tell me not to say the words conservative Republicans don't want to hear, like "rights" or "justice," and that pointed out the vacuity of the phrase "equality before the law," since the law prevents rich and poor alike from sleeping under bridges. But the reply did not stop there; it went on to point out that if a pregnant woman were to make a decision on whether to carry the fetus to term on the basis of whether the child, when born, could hope to achieve the self-sufficiency Republicans recommend for the disabled, conservatives would be the first people in her face, blocking that decision by invoking the "rights" of the unborn.

I want to point out that this little exercise in rhetorical analysis and critical legal studies was undertaken not by a cultural studies theorist, nor by someone dependent on the knowledge industry run by bourgeois sellouts like me, but by an ordinary citizen of these United States, operating in extraordinary circumstances not of her own making. But more important, I want to pass along to you what this exchange has taught me: first, that sometimes, the cost of selling out to the discourses of policy makers is too steep to bear, particularly if it means disavowing the languages of social and cultural justice; and second, that Ms. Curtis's reply, mixing righteous indignation with a keen eye for ideological contradiction, should serve to remind us that our task in selling out is not to capitulate to the terms our historical moment has offered us, but rather to find the terms with which we can best *contest* those terms, and in so doing redescribe and redefine both our cultural politics and our social policies.

NOTES

1. See Pinker and Benhabib et al.
2. As the next few pages indicate, my primary concern was that I would be faulted for having sold my subjects short in the pages of a prestigious literary magazine, and that black and nonblack critics alike would thereby read the harshest moments of my review as evidence that the *New Yorker* would take up such subjects only to dismiss them. Imagine my surprise, then, when in the spring and summer of 1995,

as the "new black intellectuals" suddenly became the topic *du jour* in every major American forum of letters, my article was singled out as an instance of mere "puffery" and "celebrity-mongering" (Wilentz) by writer after writer churning out ever shabbier essays whose only intellectual substance lay in the charge that *other* writers on the subject have lacked intellectual substance. To date the subject has seen a great deal of rhetorical posturing and turf-claiming, but as yet, only two essays (Boynton and Rivers) that touch on the question with which my essay concluded, namely, what is the role of the left public intellectual at a time when the idea of "the public" is nearly unthinkable in national public policy? (It is worth noting, for the record, that Robert Boynton wrote his essay almost a year before I wrote mine, but the *Atlantic* sat on it for quite some time, evidently secure in the belief that nothing important would happen in the world of black intellectuals between 1993 and 1995. Moreover, when the *New Yorker* ran my essay in January 1995, the *Atlantic* reportedly considered pulling Boynton's essay from the March issue—even though it was three times longer than mine and focused on different figures—as if one major essay on black intellectuals would be sufficient for the year, and two would constitute a glut.)

3. For obvious reasons, the conditions for speaking as "a Jewish intellectual" in the United States were radically different after World War II than before. Compared to the prewar *Partisan Review* writers, the postwar group was markedly more willing—indeed, sometimes compelled—to speak and write as "Jewish writers." Particularly moving in this regard was Alfred Kazin's painful reassessment of the role of the Jewish intellectual in the wake of the Holocaust, a reassessment that began with the essay "In Every Voice, in Every Ban" in the *New Republic*. However mistaken Kazin may have been to indict his fellow New York intellectuals for "our silent complicity in the massacre of the Jews," the point remains that no one writer (or group of writers) can dictate when it is proper or necessary to speak "from" or "for" one's ethnic identity and when it is preferable to speak as a "cosmopolitan." See A. Bloom, *Prodigal Sons,* esp. 137–39.

4. See Pattullo. Though Pattullo's odious essay attracted a great deal of criticism from *Commentary*'s less rabid readers, the magazine itself has not retreated an inch, calling Pattullo's essay (in a promotional mailing) "the most carefully reasoned argument for maintaining society's preference for heterosexuality over homosexuality in the teaching of children."

5. The idea that the "racial justice" provision would create a quota for death sentences is almost too bizarre to admit of discussion, but one thing about this argument seems noteworthy: even though every conservative pundit from George Will to Cal Thomas denounced the provision, none of them realized (or cared to admit) that the provision would not mandate the execution of more *white* prisoners. In other words, the provision would simply have asked the courts to consider whether blacks were being executed in wildly disproportionate numbers—and, if so, to take that disproportion into consideration in sentencing. The provision might have prevented a handful of black inmates from being executed, perhaps, but it certainly would not have increased the number of white inmates executed. The fact that the media's conservative white chorus screamed so loud and so long over a provision that merely might have lowered (slightly) the number of black inmates executed seems to me one plausible explanation for why white and black Americans have such radically different perceptions of how the nation's judicial system works.

6. See Koenig.

CULTURAL CRITICISM AND THE POLITICS OF SELLING OUT

WORKS CITED

Ahmad, Aijaz. *In Theory: Classes, Nations, Literatures.* London: Verso, 1992.

Aimone, Joseph. "The Dismantling of Higher Education." Letter. *Council Chronicle of the National Council of Teachers of English,* September 1995, 6.

Amerika, Mark. *The Kafka Chronicles.* Boulder, CO, and Normal, IL: Fiction Collective 2, 1993.

Anderson, Amanda. "Cryptonormativism and Double Gestures: Reconceiving Poststructuralist Social Theory." *Cultural Critique* 21 (1992): 63–95.

Appel, Rosaire. *transiT.* Boulder, CO, and Normal, IL: Fiction Collective 2, 1993.

Barone, Dennis. "What's in a Name? The Dalkey Archive Press." *Critique* 37.3 (1996): 222–39.

Benhabib, Seyla, Judith Butler, Drucilla Cornell, and Nancy Fraser. *Feminist Contentions: A Philosophical Exchange.* New York: Routledge, 1995.

Bennett, Tony. *Outside Literature.* New York: Routledge, 1990.

——. "Putting Policy into Cultural Studies." In *Cultural Studies,* ed. Lawrence Grossberg, Cary Nelson, and Paula Treichler, 23–37. New York: Routledge, 1992.

——. "Being 'In the True' of Cultural Studies." *Southern Review* 26.2 (1993): 217–38.

Bernard, Kenneth. *From the District File.* Boulder, CO, and Normal, IL: Fiction Collective 2, 1992.

Bérubé, Michael. *Public Access: Literary Theory and American Cultural Politics.* London: Verso, 1994.

——. "Life As We Know It: A Father, a Son, and Genetic Destiny." *Harper's* 289 (December 1994): 41–51.

——. "Public Academy." *New Yorker,* 9 Jan. 1995, 73–80.

——. *Life As We Know It: A Father, a Family, and an Exceptional Child.* New York: Pantheon, 1996.

——. "The Abuses of the University." *American Literary History,* forthcoming.

Bérubé, Michael, and Cary Nelson. *Higher Education under Fire: Politics, Economics, and the Crisis of the Humanities.* New York: Routledge, 1995.

Bloom, Alexander. *Prodigal Sons: The New York Intellectuals and Their World.* New York: Oxford University Press, 1986.

Bloom, Alexander, and Wini Breines. *Takin' It to the Streets: A Sixties Reader.* New York: Oxford University Press, 1995.

Bloom, Harold. *The Western Canon: The Books and School of the Ages.* New York: Harcourt Brace, 1994.

Boggs, Carl. *Intellectuals and the Crisis of Modernity.* Albany: State University of New York Press, 1993.

Boynton, Robert S. "The New Intellectuals." *Atlantic Monthly,* March 1995, 53–70.

Bromwich, David. *Politics by Other Means: Higher Education and Group Thinking.* New Haven: Yale University Press, 1992.

Brooks, Peter. "Aesthetics and Ideology—What Happened to Poetics?" In Levine 153–67.

Brustein, Robert. Letter. *New Yorker,* 30 Jan. 1995, 10.

Cain, William. "A Literary Approach to Literature: Why English Departments Should Focus on Close Reading, Not Cultural Studies." *Chronicle of Higher Education,* 13 Dec. 1996, B4–B5.

Carlberg, Conrad, and Kenneth Kavale. "The Efficacy of Special versus

Regular Class Placements for Exceptional Children: A Meta-Analysis." *Journal of Special Education* 14 (1980): 295–309.

Castañeda, Omar S. *Remembering to Say "Mouth" or "Face."* Boulder, CO, and Normal, IL: Fiction Collective 2, 1993.

Cheney, Lynne. *Telling the Truth: Why Our Culture and Our Country Have Stopped Making Sense, and What We Can Do about It.* New York: Simon and Schuster, 1995.

Cleaveland, Bradford. "A Letter to Undergraduates." In Lipset and Wolin 66–81.

Connor, Steven. *Postmodernist Culture: An Introduction to Theories of the Contemporary.* Oxford: Basil Blackwell, 1989.

Crimp, Douglas, ed. *AIDS: Cultural Analysis/Cultural Activism.* Cambridge: MIT Press, 1988.

Cruz, Ricardo Cortez. *Straight outta Compton.* Boulder, CO, and Normal, IL: Fiction Collective 2, 1992.

Dasenbrock, Reed Way. "What to Teach When the Canon Closes Down: Toward a New Essentialism." In *Reorientations: Critical Theories and Pedagogies,* ed. Bruce Henricksen and Thaïs E. Morgan. Urbana: University of Illinois Press, 1990.

Davidson, Carl. "The New Radicals in the Multiversity: An Analysis and Strategy for the Student Movement." In Katope and Zolbrod 273–96.

D'Souza, Dinesh. *The End of Racism: Principles for a Multiracial Society.* New York: Free Press, 1995.

———. *Falwell: Before the Millennium.* Washington, DC: Regnery Gateway, 1984.

Dugger, Celia W. "U.S. Says Mental Impairment Might Be a Bar to Citizenship." *New York Times,* 19 Mar. 1997, A1, A12.

Dyson, Michael Eric. *Making Malcolm: The Myth and Meaning of Malcolm X.* New York: Oxford University Press, 1994.

Eagleton, Terry. *Literary Theory: An Introduction.* Minneapolis: University of Minnesota Press, 1983.

Eakin, Emily. "Walking the Line." *Lingua Franca* 6.3 (March–April 1996): 52–60.

Easthope, Antony. *Literary into Cultural Studies.* New York: Routledge, 1991.

Eden, Kathy. *Hermeneutics and the Rhetorical Tradition: Chapters in the Ancient Legacy and Its Humanist Reception.* New Haven: Yale University Press, 1997.

Editorial, *New Criterion* 14.6 (February 1996): 2–3.

Eurudice. *F/32.* Boulder, CO, and Normal, IL: Fiction Collective 2, 1990.

Fish, Stanley. *Doing What Comes Naturally: Change, Rhetoric, and the Practice of Theory in Literary and Legal Studies.* Durham, NC: Duke University Press, 1989.

———. *Professional Correctness: Literary Studies and Political Change.* New York: Oxford University Press, 1995.

Fowler, Cedric. "The Crisis in the Ph.D." *New Outlook,* June 1933, 39–42.

Franklin, Phyllis, and Sandra M. Gilbert. Letter. MLA Booklet, 9 Feb. 1996, 1–2.

Free Speech Movement. *FSM: Moral Impetus, the Factory, and the Society.* In Lipset and Wolin 209–16.

Frye, Northrop. *Anatomy of Criticism.* Princeton: Princeton University Press, 1957.

Fuchs, Douglas, and Lynn S. Fuchs. "What's 'Special' about Special Education?" *Phi Delta Kappan,* March 1995, 522–30.

Gadamer, Hans-Georg. *Truth and Method.* 2d ed. Trans. Garrett Barden and John Cumming. New York: Continuum, 1975.

Gage, Beverly. "Have You No Shame?" *New Haven Advocate,* 21 Dec. 1995, 11.

Genovese, Eugene. "Living with Inequality." Symposium on *The Bell Curve. National Review,* 5 Dec. 1994, 44–46.

Gilman, Sander. "Jobs: What We (Not They) Can Do." *MLA Newsletter* 27.4 (1995): 4–5.

Graff, Gerald. *Professing Literature: An Institutional History.* Chicago: University of Chicago Press, 1987.

———. Foreword to *Universities and the Myth of Cultural Decline,* by Herron, 9–19.

Greenberg, Daniel S. "So Many Ph.D.s." *Washington Post,* 2 Jul. 1995, C7.

Greenhouse, Steven. "Labor Board Plans a Suit against Yale." *New York Times,* 19 Nov. 1996, B6.

Gross, Paul R., and Norman Levitt. *Higher Superstition: The Academic Left and Its Quarrels with Science.* Baltimore: Johns Hopkins University Press, 1994.

Grossberg, Lawrence. *We Gotta Get out of This Place: Popular Conservatism and Postmodern Culture.* New York: Routledge, 1992.

Grossman, Richard. *The Alphabet Man.* Boulder, CO, and Normal, IL: Fiction Collective 2, 1993.

Guillory, John. *Cultural Capital: The Problem of Literary Canon Formation.* Chicago: University of Chicago Press, 1993.

Harris, Charles B. Introduction. *Critique* 37.3 (1996): 163–70. Special issue on independent presses and contemporary American literature.

Henry, William H., III. *In Defense of Elitism*. New York: Doubleday, 1994.

Herron, Jerry. *Universities and the Myth of Cultural Decline*. Detroit: Wayne State University Press, 1988.

Hohendahl, Peter Uwe. *The Institution of Criticism*. Ithaca: Cornell University Press, 1982.

Homans, Margaret. Letter to Phyllis Franklin. MLA Booklet, 9 Feb. 1996, 10–11.

Hook, Sidney. *Academic Freedom and Academic Anarchy*. New York: Cowles, 1970.

"Hot Type." *Chronicle of Higher Education*, 14 Dec. 1994, 10.

Huber, Bettina J. "The MLA's 1991–92 Survey of Ph.D. Placement: The Latest English Findings and Trends through Time." *ADE Bulletin* 108 (Fall 1994): 42–51.

Hunter, Ian. *Culture and Government: The Emergence of Literary Education*. London: Macmillan, 1988.

———. "Setting Limits to Culture." *New Formations* 4 (1988): 103–23.

Jameson, Fredric. "On 'Cultural Studies.'" *Social Text* 34 (1993): 17–52.

Jauss, Hans-Robert. *Toward an Aesthetic of Reception*. Trans. Timothy Bahti. Minneapolis: University of Minnesota Press, 1982.

———. *Question and Answer: Forms of Dialogic Understanding*. Trans. Michael Hays. Minneapolis: University of Minnesota Press, 1989.

Johnson, Bayard. *Damned Right*. Boulder, CO, and Normal, IL: Fiction Collective 2, 1994.

Katope, Christopher, and Paul Zolbrod. *The Rhetoric of Revolution*. New York: Macmillan, 1970.

Kernan, Alvin. *The Death of Literature*. New Haven: Yale University Press, 1990.

Kerr, Clark. *The Uses of the University*. Cambridge: Harvard University Press, 1963.

Koenig, Robert L. "New Study Backs Danforth on Civil Rights Bill." *St. Louis Post-Dispatch*, 29 Jul. 1991, B1, B4.

Langbauer, Laurie. "The Celebrity Economy of Cultural Studies." *Victorian Studies* 36.4 (1993): 466–72.

Law, Jules David. "Uncertain Grounds: Wittgenstein's *On Certainty* and the New Literary Pragmatism." *New Literary History* 19.2 (1988): 319–36.

Levine, George. "The Real Trouble." *Profession 93* (1993): 43–45.

———, ed. *Aesthetics and Ideology*. New Brunswick, NJ: Rutgers University Press, 1994.

Lewis, Philip. *Life of Death*. Boulder, CO, and Normal, IL: Fiction Collective 2, 1993.

Leyner, Mark. *I Smell Esther Williams and Other Stories.* Boulder, CO, and Normal, IL: Fiction Collective 2, 1983. (Reprint originally published by Fiction Collective.)

Limbaugh, Rush. *See, I Told You So.* New York: Simon and Schuster, 1993.

Lipset, Seymour Martin, and Sheldon S. Wolin, eds. *The Berkeley Student Revolt: Facts and Interpretations.* New York: Doubleday, 1965.

Lott, Eric. *Love and Theft: Blackface Minstrelsy and the American Working Class.* New York: Oxford University Press, 1993.

———. "Cornel West in the Hour of Chaos: Culture and Politics in *Race Matters.*" *Social Text* 40 (1994): 39–50.

Luttwak, Samuel. "Why Fascism Is the Wave of the Future." *London Review of Books,* 7 Apr. 1994, 3, 6.

Lyon, Janet. "Transforming Manifestoes: A Second-Wave Problematic." *Yale Journal of Criticism* 5.1 (1991): 101–27.

Lyotard, Jean-François. *The Postmodern Condition: A Report on Knowledge.* Trans. Geoff Bennington and Brian Massumi. Theory and History of Literature, vol. 10. Minneapolis: University of Minnesota Press, 1984.

MacDonald, Heather. "Why Johnny Can't Write." *Public Interest* 120 (1995): 3–13.

Mazza, Cris. *Animal Acts.* Boulder, CO, and Normal, IL: Fiction Collective 2, 1988.

———. *Is It Sexual Harassment Yet?* Boulder, CO, and Normal, IL: Fiction Collective 2, 1991.

———. *Revelation Countdown.* With images by Ted Orland. Boulder, CO, and Normal, IL: Fiction Collective 2, 1993.

Mazza, Cris, Jeffrey DeShell, and Elisabeth Sheffield, eds. *Chick-Lit 2: No Chick Vics.* Boulder, CO, and Normal, IL: Fiction Collective 2, 1996.

McCaffery, Larry, ed. *Avant-Pop: Fiction for a Daydream Nation.* Boulder, CO, and Normal, IL: Fiction Collective 2, 1993.

McLaughlin, Robert L. "Oppositional Aesthetics/Oppositional Ideologies: A Brief Cultural History of Alternative Publishing in the United States." *Critique* 37.3 (1996): 188–204.

Messer-Davidow, Ellen, David R. Shumway, and David J. Sylvan, eds. *Knowledges: Historical and Cultural Studies in Disciplinarity.* Charlottesville: University of Virginia Press, 1993.

Meyer, Stephen. "The Role of Scientists in the 'New Politics.'" *Chronicle of Higher Education,* 26 May 1995, B1–B2.

Mitchell, W. J. T. "Pluralism as Dogmatism." *Critical Inquiry* 12.3 (1986): 494–502.

Morton, Donald. "The Politics of Queer Theory in the (Post)Modern Moment." *Genders* 17 (1993): 121–50.

Muller, Jerry Z. "A Conservative Defense of the Humanities Endowment." *Wall Street Journal,* 12 Apr. 1995, A14.

Nanterre Manifesto. Trans. Peter Brooks. In Katope and Zolbrod 268–73.

Nelson, Cary. "Lessons from the Job Wars: Late Capitalism Comes to Campus." *Social Text* 44 (1995): 119–34.

———. "Lessons from the Job Wars: What Is to Be Done?" *Academe: Bulletin of the American Association of University Professors* 81.6 (November-December 1995): 18–25.

Nelson, Cary, and Michael Bérubé. "Graduate Education Is Losing Its Moral Base." *Chronicle of Higher Education,* 17 Mar. 1994, B1–B3.

Nielson, Jim, and Gregory Meyerson. "Public Access Limited." Review of *Public Access: Literary Theory and American Cultural Politics,* by Michael Bérubé. *minnesota review* 46–47 (1996): 263–73.

Oakley, Francis. "Against Nostalgia: Reflections on Our Present Discontent in American Higher Education." In *The Politics of Liberal Education,* ed. Darryl J. Gless and Barbara Herrnstein Smith. Durham, NC: Duke University Press, 1992.

O'Hara, Daniel T. "Lentricchia's Frankness." *boundary 2* 21.2 (1994): 40–62.

Ohmann, Richard. *English in America: A Radical View of the Profession.* New York: Oxford University Press, 1976.

———. *Politics of Letters.* Middletown, CT: Wesleyan University Press, 1987.

Patai, Daphne, and Noretta Koertge. *Professing Feminism: Cautionary Tales from the Strange World of Women's Studies.* New York: Basic Books, 1994.

Patterson, Annabel. Letter to Phyllis Franklin. MLA Booklet, 9 Feb. 1996, 5–7.

Pattullo, E. L. "Straight Talk about Gays." *Commentary* 94.6 (1992): 21–24.

Pell, Derek. "The Elements of Style." In McCaffery 49–81.

Pinker, Steven. *The Language Instinct.* New York: Morrow, 1994.

Posnock, Ross. "A View from an English Department." Symposium on "Intellectual History in the Age of Cultural Studies." *Intellectual History Newsletter* 18 (1996): 18–20.

Quartermain, Peter. "Trusting the Reader." *Chicago Review* 32.2 (1980): 65–74.

Quayle, Dan. *Standing Firm: A Vice-Presidential Memoir.* New York: HarperCollins, 1994.

Readings, Bill. *The University in Ruins.* Cambridge: Harvard University Press, 1996.

Rich, Adrienne. "Vesuvius at Home: The Power of Emily Dickinson." In *On Lies, Secrets, and Silence: Selected Prose, 1966–78.* New York: Norton, 1979.

Rivers, Eugene F. "Beyond the Nationalism of Fools: Toward an Agenda for Black Intellectuals." *Boston Review* 20.3 (1995): 16–18.

Robbins, Bruce. *Secular Vocations: Intellectuals, Professionalism, Culture.* London: Verso, 1993.

———. " 'Real Politics' and the Canon Debate." *Contemporary Literature* 35.2 (1994): 365–75.

Robinson, Lou. *Napoleon's Mare.* Boulder, CO, and Normal, IL: Fiction Collective 2, 1991.

Rooney, Ellen. "Discipline and Vanish: Feminism, the Resistance to Theory, and the Politics of Cultural Studies." *differences* 2.3 (1990): 14–28.

Ross, Andrew. "Green Ideas Sleep Furiously: An Interview with Andrew Ross." By Mark McGurl. *Lingua Franca* 5.1 (1994): 57–65.

———. "Undisciplined: An Interview with Andrew Ross." By Jeffrey Williams and Mike Hill. *minnesota review* 45–46 (1996): 77–94.

Said, Edward W. "Opponents, Audiences, Constituencies, and Community." *Critical Inquiry* 9.1 (1982): 1–26.

———. *Representations of the Intellectual.* New York: Pantheon, 1994.

Savio, Mario. "An End to History." In Lipset and Wolin 216–19.

Servin, Jacques. *Mermaids for Attila.* Boulder, CO, and Normal, IL: Fiction Collective 2, 1991.

Shumway, David. *Creating American Civilization: A Genealogy of American Literature as an Academic Discipline.* Minneapolis: University of Minnesota Press, 1994.

Simpson, David. *The Academic Postmodern and the Rule of Literature: A Report on Half-Knowledge.* Chicago: University of Chicago Press, 1995.

Smith, Alison T. "Secondary Education: Still an Ignored Market." *Profession 96* (1996): 69–72.

Solomon, Robert, and Jon Solomon. *Up the University: Re-Creating Higher Education in America.* Reading, MA: Addison-Wesley, 1993.

Sowell, Thomas. *Race and Culture: A World View.* New York: Basic Books, 1994.

Spacks, Patricia Meyer. "The Academic Marketplace: Who Pays Its Costs?" *MLA Newsletter* 26.2 (1994): 3.

Starr, Peter. *Logics of Failed Revolt: French Theory after May '68.* Stanford: Stanford University Press, 1995.

Steiner, Robert. *Broadway Melody of 1999*. Boulder, CO, and Normal, IL: Fiction Collective 2, 1993.

Students for a Democratic Society. *The Port Huron Statement*. In Bloom and Breines 61–74.

Stuefloten, D. N. *Maya*. Boulder, CO, and Normal, IL: Fiction Collective 2, 1992.

Tarnawsky, Yuriy. *Three Blondes and Death*. Boulder, CO, and Normal, IL: Fiction Collective 2, 1993.

Thompson, Kathryn. *Close Your Eyes and Think of Dublin: Portrait of a Girl*. Boulder, CO, and Normal, IL: Fiction Collective 2, 1991.

Tomasky, Michael. "Left for Dead." *Village Voice*, 22 Nov. 1994, 1, 19–20.

Trent, James. *Inventing the Feeble Mind: A History of Mental Retardation in the United States*. Berkeley: University of California Press, 1994.

Turner, Darwin. *In a Minor Chord: Three Afro-American Writers and Their Search for Identity*. Carbondale: Southern Illinois University Press, 1971.

Urgo, Joseph. "An Obscure Destiny, This Business of Teaching English." *Profession 96* (1996): 134–38.

Villa, Richard A., Susan Stainback, William Stainback, and Jacqueline Thousand, eds. *Restructuring for Caring and Effective Education: An Administrative Guide to Creating Heterogeneous Schools*. Brookline, MA: Paul H. Brookes, 1992.

Vizenor, Gerald. *Griever: An American Monkey King in China*. New York and Boulder, CO: Fiction Collective, 1987.

Watt, Stephen. "The Human Costs of Graduate Education; or, The Need to Get Practical." *Academe: Bulletin of the American Association of University Professors* 81.6 (November-December 1995): 30–35.

Weisberg, Jacob. "Who, Me? Prejudiced?" *New York*, 17 Oct. 1994, 26–30.

Wilentz, Sean. "Race, Celebrity, and the Intellectuals." *Dissent* 42.3 (1995): 293–99.

Will, George. "Teach Johnny to Write." *Washington Post*, 2 Jul. 1995, C7.

Young, Cynthia. "On Strike at Yale." *minnesota review* 45–46 (1996): 179–95.

Zavarzadeh, Mas'ud. " 'The Stupidity That Consumption Is Just as Productive as Production': In the Shopping Mall of the Post-al Left." *College Literature* 21.3 (1994): 92–114.

Apple, Michael, 165–66
Association of Literary Critics and Scholars, 12
Autobiography of an Ex-Colored Man (J. W. Johnson), 175
Avant Pop, 124
Avant-Pop fiction, 120–24

Baker, Houston, 209
Barnes, Djuna, 137
Barthes, Roland, 124
Bell, Derrick, 230
Bellow, Adam, 213
Benda, Julien, 222
Bennett, Tony, 163, 229, 232
Bennett, William, 21
Bernard, Kenneth, 128, 134
Bérubé, James, 173, 219–20
Bloom, Harold: and cultural relativism, 91–93; and decline of literary studies, 26–27; and value of literary studies, 109
Boas, Franz, 208–9
Boynton, Robert, 241
Bromwich, David, 93–97
Brooks, Peter, 35, 42, 48, 51
Brustein, Robert, 109
Buchanan, Pat, 213–14
Bush, George, 235
Butler, Judith, 150

Cain, William, 5–9
Carew, Thomas, 48
Carville, James, 228
Castañeda, Omar, 129–30
Celebrity, cultural studies and, 99–100
Chariot of Wrath, The: The Message of John Milton to Democracy at War (Knight), 150–51
Chavez, Linda, 214
Cheney, Lynne, 180, 213
City University of New York, 77
Civil Rights Act of 1964, 210–11, 213
Civil Rights Act of 1991, 235–36
Clinton, Bill, 228–29
Close Your Eyes and Think of Dublin: Portrait of a Girl (Thompson), 127–28
Cole, Johnnetta, 209
Collier, Peter, 226
Congressional Black Caucus, 228–29
Conservatism, 211–15. *See also* New Right
Crimp, Douglas, 220
Critical legal studies, 151–52

Critical thinking: existence of, 155; reversibility of, 158
Cruz, Ricardo Cortez, 124
Cultural capital: and critical thinking, 158; discriminating function of, 23–24; and future of literary study, 26; literary study as, 19; and productive capital, 23
Cultural Capital: The Problem of Literary Canon Formation (Guillory), 19, 23
Cultural politics, 227–29, 231, 236
Cultural relativism, 208–9
Cultural Studies (Grossberg, Nelson, and Treichler, eds.), 99, 110
Cultural studies: advent of, 4; "celebrity economy" of, 99–100; and conflict with literary study, 4–6, 12–18, 26–27, 105–6, 108–9; and contemporary writing in English, 28; controversy over, 6; crisis of evaluation in, 93; and English departments, 25, 27; future of, 27–28, 110; interdisciplinary nature of, 106; and Left, 25–26, 223–24, 226–27; and MLA, 5; and New Right, 223–24; and nonacademic constituencies, 150; and public policy, 223–24, 231–40; and public sphere, 216, 219; as response to market value of literary study, 23; and "selling out," 217–18, 236; skepticism about, 155–56; as survivalism, 25; and theory, 87; value of, 83–84
Culture and Imperialism (Said), 161
Curtis, Janet, 240

Dalkey Archive Press, 136–39
Damned Right (B. Johnson), 121–22
Danforth, John, 235
Dasenbrock, Reed Way, 28–30
Davidson, Carl, 188–89
Davis, David Brion, 48
Deconstruction, 151–52, 154
Delaporte, François, 220
Denning, Michael, 48, 53–54
de Saint-Simon, Henri, 222
Disabilities, human: and "full inclusion," 181–82; and immigration legislation, 168–69; and New Right, 239–40; representations of, 155, 219–20
Disciplinarity, 183; and academic job market, 201–3; and external constituencies, 197–97, 200; and institutional change, 184–87; and paradigm shift in cultural studies, 198–99; and postmodern university, 187–98. *See also* Interdisciplinarity

INDEX

Doctoral programs, shrinkage of, 79–85
Down syndrome: and immigration legislation, 168–69; and New Right, 239–40;
  representations of, 150, 155, 219–20
D'Souza, Dinesh, 204; and cultural relativism, 208–9; and extreme conservatism,
  213–15; and "liberal antiracism," 208;
  and proposed repeal of Civil Rights Act of
  1964, 210–11, 213; on slavery, 207, 208;
  on stereotypes, 205–6; success of, 211–12
Du Bois, W. E. B., 209
Dyson, Michael Eric, 230–31

Eagleton, Terry, 29, 96
Eakin, Emily, 48, 51
Early retirement, of faculty, 73–79
Employment, academic: crisis in, 68–69;
  definition of, 37–38; and hiring practices,
  71–72; and theory, 86–87. *See also* Job
  market
End of Intelligent Writing, The (Kostelanetz),
  115
End of Racism, The (D'Souza): 204–15
English departments: and canon revision,
  29–30; and cultural studies, 25, 27; disciplinarity of, 183–203; functions of, 111;
  future of, 27–28, 107; and graduate student labor, 87; and institutional capital of
  English, 21; interdisciplinarity of, 106–8;
  and job market, 22–23, 80, 81; interest of
  undergraduates in, 21–22; multicultural
  curricula in, 21; quality of new hires in,
  6–9; reactionaries in, 109; and Responsibility-Centered Management, 33–34; self-
  descriptions of, 9–11, 25; size of, 20; and
  writing, teaching of, 32, 108, 111
English in America (Ohmann), 101
Epstein, Richard, 169, 210
Eskin, Blake, 233
Eurudice, 123–24

F/32 (Eurudice), 123–24, 140
Faculty: alienation of senior from junior,
  103; as celebrities, 99; and conservatives,
  persecution of, 211–12; downsizing of,
  68; and early retirement, 73–79; and elimination of tenure, 76–77; and graduate student unionization, 38, 41–49, 51; job requirements for, 74–75; and knowledge,
  state of, 190; and need to cut doctoral programs, 79–85; obligations of, 73–85; part-
  time, 62–63, 89; political advocacy by,

170–82; and publishing, 102–4; reaction-
  ary, 109; as replacement workers, 46–47,
  56; and Yale graduate student strike, 39–
  41, 48
Fairbanks, Lauren, 137
Falwell: Before the Millennium (D'Souza),
  204–5
Federation of University Employees, 39
Feminist theory, as advocacy, 176–77
Fiction Collective, 115–16, 118
Fiction Collective 2 (FC2), 113–21, 133–40
Finder, Henry, 221
Fish, Stanley: and critical thinking, 155–56;
  and impact of literary criticism, 145–55,
  169; and institutional change, 184–87;
  and metaphysics of stasis, 157; and profes-
  sionalism of literary study, 159–60
Ford Foundation, 176, 177
Foucault, Michel, 11
Fowler, Cedric, 68
Franklin, Phyllis, 42, 56
Free Speech Movement, Berkeley, 188–89
From the District File (Bernard), 128
Frye, Northrop, 16, 91, 159

Gadamer, Hans-Georg, 171–72
Genetic foundationalism, 219–20
Genovese, Eugene, 179, 207
Gerontion Effect, 95
Gilman, Sander, 58–62
Goldwater, Barry, 214
Grade strike, Yale, 39–41, 48–50
Graduate Employees Organization (GEO),
  University of Illinois: and dependence on
  graduate student instructors, 83; faculty
  support of, 66. *See also* Unionization of
  graduate students
Graduate Employees and Students Organization (GESO), Yale: affiliation with nonacademic unions of, 43–44, 54; faculty response to, 38, 41–48, 51, 53–56, 62–63;
  and faculty-student relationship, 39; and
  grade strike, 39–41; and MLA, 42–43,
  55–58, 62; and NLRB ruling, 40, 63; support for, at other institutions, 49. *See also*
  Unionization of graduate students
Graduate students: and adjunct faculty, 44–
  45; compensation of, 45, 47, 87–89; demands on, 102–3; and early retirement of
  faculty, 73–74; as employees, 37–39; exploitation of, 49–50, 53–54, 87–89; and
  high-school teaching, 84–86; and Master

INDEX

North Point, 137
Novel and the Police, The (Miller), 16–17

Oakley, Francis, 22
"Objectivity," 171–72
O'Brien, John, 138, 139
O'Hara, Daniel, 100
Ohmann, Richard, 20, 101, 111

Paton, Diana, 48
Patterson, Annabel, 42–43, 47–49, 55
Pell, Derek, 124
Peer reviews, 97
Pinker, Steven, 220
Pisan Cantos (Pound), 17
Podhoretz, Norman, 225
Political advocacy, by faculty, 170–82
"Political correctness," 170–71, 234–36
Political Unconscious, The (Frye), 16
Politics of Letters (Ohmann), 20
Politics by Other Means (Bromwich), 93–95
Posnock, Ross, 7–8
Postdoctoral mentored teaching fellowships, 58–61
Post-tenure review, 77
Pound, Ezra, 17–18
Professional Correctness: Literary Studies and Political Change (Fish), 145–50, 157, 159
Professionalism, 69; and distinctiveness, 159; and early retirement, 74–77; and graduate student unions, 89; and political advocacy, 178–81
Professional organizations, role of, 69–70
Proposition 187, 238
"Public": attacks on, by New Right, 218–19, 237–38; conception of, by intellectuals, 225
Public Access: Literary Theory and American Cultural Politics (Bérubé), 144, 145, 148–49, 150
Public spheres, 216
Publishing, necessity of, for faculty, 102–4

Quartermain, Peter, 119
Quayle, Dan, 92, 214

Race and Culture: A World View (Sowell), 223, 232–33
Race Matters (West), 221
Racism: and extreme conservatism, 215; and "liberal antiracism," 205, 208; and proposed repeal of Civil Rights Act of 1964, 210–11, 13. See also *End of Racism, The*

Radway, Janice, 5
Rahv, Philip, 225
"Rational discrimination," 205–6
Reagan, Ronald, 152–53
Remembering to Say "Mouth" or "Face" (Castañeda), 129–30
Responsibility-Centered Management, 33–34
Revelation Countdown (Mazza), 130–31
Rich, Adrienne, 175
Right, political. *See* New Right
Robbins, Bruce, 24
Robinson, Lou, 124, 125
Rockefeller Foundation, 176, 177
Rooney, Ellen, 106
Rorty, Richard, 212
Rose, Tricia, 231
Rosenberg, Milton, 23
Ross, Andrew, 168, 224, 231, 234

Said, Edward: and audience for cultural criticism, 218; and political orientation of cultural studies, 6, 15, 161, 231; and representations of the intellectual, 222
Savio, Mario, 188
Secondary education. *See* High-school teaching
Sedgwick, Eve Kosofsky, 15–16
See, I Told You So (Limbaugh), 230
"Selling out," 217–18, 220–22
Servin, Jacques, 129
Shaw, E. Clay, 169
Sheehan, Aurelie, 136
Simpson, David, 25
Simpsons, The, 227–28
Smith, Alison T., 85
Solomon, Jon, 106–7
Solomon, Robert, 106–7
Sorrentino, Christopher, 137
Sorrentino, Gilbert, 138
Sowell, Thomas, 92, 223, 232–34
Spacks, Patricia Meyer, 62
Standing Firm (Quayle), 92
Steiner, Wendy, 6
Straight outta Compton (Cruz), 124
Strauss, Leo, 152–53
Stuefloten, D. N., 126–27
Sukenick, Ron, 116, 118, 136
Suleri-Goodyear, Sara, 48

Tarnawsky, Yuriy, 124–25
Tenure: and academic reform, 66; and aes-

ABOUT THE AUTHOR

MICHAEL BÉRUBÉ is a professor of English at the University of Illinois at Urbana-Champaign and Director of the Illinois Program for Research in the Humanities. Born in New York City in 1961, he received his B.A. in English from Columbia University in 1982 and his M.A. and Ph.D. from the University of Virginia in 1986 and 1989, respectively.

His previous books are *Marginal Forces/Cultural Centers: Tolson, Pynchon, and the Politics of the Canon* (1992), *Public Access: Literary Theory and American Cultural Politics* (1994), *Higher Education under Fire: Politics, Economics, and the Crisis of the Humanities* (1995), coedited with Cary Nelson, and *Life As We Know It: A Father, a Family, and an Exceptional Child* (1996).

When he's not reading or writing, he's playing drums with his rock band, Nastybake, or haunting the local roller hockey leagues. He lives in Champaign, Illinois, with Janet Lyon and their two sons, Nicholas, 11, and James, 6.

About the Illustrator

UTHMAN GUADALUPE is an aspiring writer, filmmaker, and illustrator based in Maryland. He is passionate about drawing and specializes in traditional and digital art, manga, comics, and cartoons. Uthman has illustrated other children's books, including The First Day of Ramadan, The Secret of My Hijab, and Why Do Muslims...? 25 Questions for Curious Kids.

Made in the USA
Middletown, DE
06 May 2021

39132901R00047